Carlos Mateus Rodrigues Amorim
Bruna R. Furtado
Alberto C. Melo Lima

Comparison of thermal behaviour: conventional roof and eco-tile

Carlos Mateus Rodrigues Amorim
Bruna R. Furtado
Alberto C. Melo Lima

Comparison of thermal behaviour: conventional roof and eco-tile

Eco-roof and its thermal behaviour

ScienciaScripts

Imprint
Any brand names and product names mentioned in this book are subject to trademark, brand or patent protection and are trademarks or registered trademarks of their respective holders. The use of brand names, product names, common names, trade names, product descriptions etc. even without a particular marking in this work is in no way to be construed to mean that such names may be regarded as unrestricted in respect of trademark and brand protection legislation and could thus be used by anyone.

Cover image: www.ingimage.com

This book is a translation from the original published under ISBN 978-613-9-65832-9.

Publisher:
Sciencia Scripts
is a trademark of
Dodo Books Indian Ocean Ltd. and OmniScriptum S.R.L publishing group

120 High Road, East Finchley, London, N2 9ED, United Kingdom
Str. Armeneasca 28/1, office 1, Chisinau MD-2012, Republic of Moldova, Europe
Printed at: see last page
ISBN: 978-620-7-96605-9

Copyright © Carlos Mateus Rodrigues Amorim, Bruna R. Furtado, Alberto C. Melo Lima
Copyright © 2024 Dodo Books Indian Ocean Ltd. and OmniScriptum S.R.L publishing group

ACKNOWLEDGEMENTS

I thank God first of all for this opportunity.

To **my family,** who have always encouraged me and given me the strength to get where I am.

And to all **my friends,** who, through their socialising and experiences together, have helped shape the person I am today.

Carlos Mateus Rodrigues Amorim

ACKNOWLEDGEMENTS

To **God,** for everything.

To my parents **João and Alzira,** for their support, trust, encouragement and presence in the most difficult times.

To Professor **Alberto Carlos de Melo Lima** for his guidance, dedication and fundamental teachings in carrying out this research.

To all the **staff and residents of the Agua Cristal condominium** for all their support throughout our data collection. A special thank you to the neighbour on the plot for all the times he unwittingly gave me water and to all the condominium guards for the countless times they saved me from the jumping frogs.

To my friend and partner **Mateus Amorim** for his friendship, patience, motivation, attention and joy throughout this work.

To Professors **Eliane de Castro Coutinho** and **Marco Valério de Albuquerque Vinagre** for agreeing to sit on the examining board.

To Mr **Francisco de Assis Sandim Nery** of INMET for helping me collect climatological data.

To my grandmother **Júlia Fernandes** for all her valuable tips on grasses.

To my brother **Breno Furtado** for his support and trust.

To my dear friends **Ivan, Lorena, Érika, Ricardo, Adalberto, Deyvisson** and **Andreíza** for making my life so much happier.

To the others who, in some way, contributed to the realisation of this work.

Bruna Ribeiro Furtado

We must have faith. There are no futile efforts if they are used in favour of the common good."

Getúlio Vargas

SUMMARY

The current climate changes that the world is undergoing are causing concern due to their intensity and the speed with which they are happening. In the urban environment, the increase in temperature is more noticeable, since it is there that the natural environment has been strongly transformed, as seen in the replacement of vegetation with buildings, concrete and asphalt. The aim of this work was to carry out a comparative study between conventional roofing and eco-roofing, which is a type of roofing with a living layer of vegetation and whose use is associated with various benefits, analysing its thermal behaviour. Among the benefits, the eco-roof is recognised for having a thermal behaviour that mitigates the effects of rigid climates, such as excess heat or cold, and this particularity of eco-roofs allows for a greater thermal balance in the environment, thus providing better thermal comfort in the buildings in which they are installed. Two measurement units were built to monitor temperature, located in the metropolitan region of Belém. One unit used a conventional roof and the other an eco-roof. A digital *datalogger* thermometer was placed in each unit to measure the temperature every 10 minutes. The results obtained showed that the use of eco-roofs has a considerable effect on the temperature of a building's roof and, consequently, its interior. There was an average temperature reduction of 1.5°C when comparing the temperature of the internal surface of the roof tiles, while the difference between the maximum temperatures of the internal surfaces is an incredible 15.5°C. When analysing the influence of different meteorological parameters on the surface temperature of the measuring units, there is a significant relationship with the outside air temperature, humidity, solar insolation, but not so much with rainfall. The two measurement units suffer similar interference from these climatic parameters, but it can be seen that they suffer these interferences with different intensities. It can therefore be concluded that the use of eco-roofs in the Amazon region makes it easier to achieve an environment that provides thermal comfort for users.
Keywords: eco-roof, thermal comfort.

SUMMARY

CHAPTER 1

INTRODUCTION

The process of urbanisation has intensified across the globe in recent decades, with a growing population in urban areas causing significant changes to the landscape and the natural environment. As a result of this urban growth, architecture and urbanism have ended up following an international standardisation, which has overlooked the characteristics and specificities of the various countries and regions within them, mainly due to the capitalist economic system of massification adopted and the form of urbanisation on a global scale, maintaining a similar constructive characteristic regardless of the location.

With the standardisation of architectural and urban planning techniques, urban centres suffer some damage considering the influence of the built environment interacting with the natural environment, such as the formation of heat islands, represented by the increase in temperature in these places, since the material used in the buildings absorbs a lot of heat, in addition to the buildings preventing the wind from circulating, there is also a significant increase in surface runoff due to soil sealing, causing floods and inundations.

The environment also suffers from the reduction in green areas, replaced by buildings and street paving. This standardisation runs counter to the main values of architecture and urbanism, since, as SIQUEIRA (2009) states, the main objective nowadays is environmental comfort, comprising the study of thermal, acoustic, light and energy conditions and the physical phenomena associated with them as one of the conditioning factors of the form and organisation of space.

With the aim of combating these harmful effects on both human beings and the environment, there are techniques within architecture and urbanism that, recognising the current demand for environmental preservation and improved quality of life, meet this precept, such as green buildings. Green buildings are a set of techniques that aim to meet individual and environmental demands against climate change on a local scale. One of the existing techniques is the use of green roofs, or ecotiles (LIMA;

BARROCA; D'OLIVEIRA, 2009). The use of green roofs is an architectural technique characterised by the use of vegetation (trees, plants, ...) on the roofs of homes and buildings, which are often used to achieve greater thermal comfort on a local scale. In addition to temperature regulation, the use of green roofs also has other benefits such as rainwater retention and aesthetic and social benefits.

The aim of this project is to study the viability of eco-roofs in buildings as a possible solution to thermal discomfort, by comparing the thermal behaviour of eco-roofs and conventional roofs. The studies in this project aim to increase knowledge of the benefits of this technique, which is not yet widely used.

CHAPTER 2

OBJECTIVES

2.1 - General Objective

A comparative study between the use of conventional and eco-roofs, with the aim of analysing the thermal behaviour of the latter and evaluating the thermal comfort obtained through its use in the Amazon region, particularly in the metropolitan area of the city of Belém, state of Pará.

2.2 - Specific objectives

- Literature review on the subject.
- Construction of two measurement units to assess thermal comfort using conventional and eco-roof tiles.
- Data acquisition in the two measurement units built over the course of a year (full cycle).
- Evaluation of the relationship between meteorological parameters and temperature through the implementation of eco-roofs in the Amazon region.

CHAPTER 3

LITERATURE REVIEW

3.1 - A Brief History

The use of green roofs dates back to ancient times when many peoples used them in different ways and with different meanings (ALMEIDA, 2008). The implementation of green roofs depends on the various characteristics of each region, but their benefits, both environmental and economic, are recognised everywhere in the world. The most famous and oldest example of this type of roof on buildings is the Hanging Gardens of Babylon, built in 450 BC at the behest of King Nebuchadnezzar (ANDRADE, 2007) and considered one of the Seven Wonders of the Ancient World. The name "Hanging Gardens" comes from the fact that the building was built in levels that gave the impression of being suspended (figure 1). Ancient reports indicate that these floors were supported by columns and had outdoor terraces landscaped with exotic trees and flowers that were irrigated by efficient hydraulic systems (ALMEIDA, 2008).

Figure 1: Hanging gardens of Babylon.
Source: Almeida, 2008.

The Vikings, who inhabited the Scandinavian Peninsula, located in the far north of Europe, used to build their houses using grass for the walls and roofs. Sometimes it was combined with mud, earth and seaweed (PECK, 1999 apud ALMEIDA, 2008). The importance of this covering for them was to protect against storms and winds, which were very strong in this region, as well as reducing heat loss in a region where winters were very harsh.

In Europe, especially in Germany, engineers, architects and designers were engaged in scientific research to perfect the technique. This scientific advance was based mainly on strong public environmental awareness combined with pressure from radical ecological groups (HENEINE, 2008). During the 1980s, the commercialisation and construction of green roofs grew by 15% to 20% per year (LOPES, 2007). Today, it is estimated that around 5-10% of roofs in Germany are green.

In the United States and Canada, there is a great deal of government incentive for building with green roofs. In New York, incentives are offered to construction companies that install green roofs on at least 50 per cent of their projects, as well as tax breaks. In the city of Chicago (known as the green roof capital of the United States), the implementation of green roofs has been encouraged since 1998, most notably Millennium Park (figure 2), with one of the largest green roofs in the world.

Figure 2: Naturalised cover of Millennium Park.
Source: Lopes, 2007.

In Latin and South America, eco-roofs are gradually gaining ground, serving as an effective solution to the problem of urban heat islands, flooding and the degradation of green spaces, which are very common in countries that have undergone irregular and accelerated occupation processes. In Mexico

there is growing acceptance among the population and the government. The Roof Policy Project, winner of the 2005 Holcim prize, which honours the best ideas for sustainable construction, proposes the construction of hanging gardens in the city of Buenos Aires with a view to "improving the conditions of individual buildings as well as environmental problems".

Green roofs are beginning to be extended to greater and lesser extents in many regions of the world, and the motivating factors for implementing the system can vary according to climate, culture and politics, as well as the results of the levels and types of incentives to promote them (HENEINE, 2008).

In Brazil, the emphasis on the application and study of the development of eco-roofs is mostly concentrated in the south and southeast regions. Studies on the benefits of the technique are being carried out at the main and most respected national universities such as UFRJ, USP, UFRGS and

UFSCAR.

Most of the projects developed are experimental, carried out on the campus of the universities themselves, with the construction of comparative prototypes (conventional roofs and eco-roofs), monitoring temperatures, surface run-off, thermal comfort and different types of grass.

2.3 - Types of conventional roofs

Azeredo (1997) divides the roof of buildings into three distinct parts: the structure, rainwater collection and the covering. The structure refers to the elements that support the roof and part of the rainwater collection system. The rainwater collection system is characterised by the elements that collect and convey the rainwater that falls on the roof. The roof, finally, is the component whose purpose is to protect the building against bad weather and which must have insulating properties.

Lee (2000) states that the main function of roofs is, in addition to weather protection, to provide adequate levels of mechanical, thermal, acoustic and visual comfort for the user. In order to achieve this, there are various options on the market for combining roofs and structures.

In Brazil, ceramic and fibre cement roof tiles are the most widely used (LONGSDON, 2002). There are also steel or aluminium roof tiles, but these are basically used on the roofs of industrial buildings. According to Azeredo (1997), the characteristics that these tiles should have are: impermeability, resistance, unalterable in terms of shape and weight, light, quick-drying, easy to lay, long-lasting, economical, easy to maintain, it should lend itself to expansion and contraction, and have good drainage.

3. 3- Definition of Eco-roofs

Green roofs or eco-roofs, the term used in this study, are characterised by the use of vegetation planted on buildings (figure 3), with a layer of soil or substrate and another layer of vegetation applied to the surface of the structure. In the literature, there are various terms to describe this different type of roof, such as living green roofs, green roofs, living roofs, green roofs, living roofs, vegetated roofs, biotiles, ecotiles (OLIVEIRA, 2009).

Figure 3: Example of a green roof.
Source: Heneine, 2008.

The choice of vegetation cover needs to be compatible with the intended implementation and management conditions. It is necessary to analyse all the existing influencing factors in order to decide on the best type of cover to use. For each region, there are different wind conditions, sunshine, temperature, rainfall and characteristic species of vegetation, which will influence the construction system adopted and maintenance (ALMEIDA, 2008).

It is also necessary to know in advance how it will be used: whether there will be access for people on the roof, whether the purpose is aesthetic, whether it will be highly visible from the surroundings and adjacent buildings, and whether there will be money available and people responsible for maintenance (KOERICK, 2010).

The literature calls green roofs with a low substrate thickness "extensive green roofs" and those with a high substrate thickness "intensive green roofs" (ROSENZWEIG et al, 2006 apud LOPES, 2007).

Other types of coverings described in the literature are semi-intensive green coverings (LOPES, 2007).

Intensive green roofs are those that require a complex structure and a great deal of maintenance. Any type of plant can be used in these systems, from grasses to large trees, as shown in figure 4. They are unlikely to be implemented on roofs with steep slopes, as the heavy weight of the substrate and vegetation contributes greatly to slipping. The intensive system is usually chosen when there is an aesthetic concern, or when certain types of plants are required (ALMEIDA, 2008).

Figure 4: Intensive green cover.

Source: Heneine, 2008.

In addition, intensive green roofs provide protection for buildings and related constructions, including long-lasting waterproofing, by forming an additional thermal layer (HENEINE, 2008).

Extensive green roofs are characterised by their simple structure and require virtually no human intervention in their management. It is a lighter system that requires little water and is more economical to implement. However, it does not usually allow trampling (figure 5) and the number of species that can be used is restricted (ALMEIDA, 2008).

Extensive roofs are also considered eco-efficient, as they contribute to environmental comfort and energy efficiency, without spending a large amount of external resources on their construction and maintenance (ALMEIDA, 2008).

Figure 5: Extensive green cover.

Source: Lopes, 2007.

Semi-intensive green roofs are generally covered with grasses, trees and small shrubs and are intermediate between extensive and intensive green roofs. They are between 12 and 20 cm thick and require some care in terms of water and nutrients and are inexpensive in terms of maintenance costs (NASCIMENTO, 2008).

Tables 1 and 2 respectively show the differences and characteristics of the types of green roofs and the advantages and disadvantages of the main types of roof, intensive and extensive.

Table 1: Types of roofs and their characteristics.

	Types of Covers		
	Extensive	Semi-intensive	Intensive
Use	Ecological (not usable)	Garden	Garden/park
Types of vegetation	Mosses/herbs/grasses	Herbs/grasses/shrubs	Grass/ perennials/ shrubs/ trees
Benefits	Water reserve/> thermal efficiency/> biodiversity	Water reserve/>thermal efficiency/ >biodiversity/use	Water reserve/>thermal efficiency/ >biodiversity/use
Substrate height	60 - 200 mm	120-250 mm	150-400 mm
Weight (saturated)	80-150 kg/m^2	120-250kg/m^2	180 - 500 kg/m^2
Type of maintenance	Nonexistent/low	Periodic	High
Need for water	No/yes	No/yes	Yes

Source: Paulo Palha (APH Magazine, no. 106).

Table 2: Description of the advantages and disadvantages between an intensive green roof and an extensive green roof.

Intensive Green Roofs	Extensive Green Roofs
Advantages	
• It provides large spaces for the establishment of a diversity of fauna and flora; • The garden can be used as a space for contemplation as well as for growing food.	• Low maintenance in terms of irrigation, pruning and fertilisation; • Low structural weight; • Greater flexibility with regard to the slope of the roof; • It does not require expertise in execution; • Suitable for implementation in existing buildings; • Relatively low cost.
Disadvantages	
• Complex and high-maintenance construction system; • High structural weight; • High initial construction and maintenance costs.	• There is usually no access to spaces for contemplation and recreation; • Limited choice of plant species.

Source: Johnston and Newton, 2004.

3.4- Construction Technologies

When using green roofs, you don't just need layers of substrate and vegetation. Green roofs are made up of several layers, each with a specific function, which can vary according to the need or location of application. These layers, which make up the green roof structure, are shown in figure 6 and are basically:

Figure 6: Overlapping layers in a green roof.
Source: Tanner & Scholz-Barth, 2004 apud Lopes, 2007.

- **Vegetation:** The choice of vegetation requires knowledge of the region's climate, the type of substrate that will be used and the resistance and maintenance requirements of the type of vegetation chosen, generally opting for those that require less maintenance.

- **Substrate:** The layer that supports the vegetation and contains the nutrients it needs. It must have good drainage and vary in thickness according to the type of vegetation used.

- **Filter layer:** This protects the drainage layer by preventing particles from the substrate from passing through and impairing drainage.

- **Drainage layer:** Responsible for removing excess water from the green roof. A porous material such as gravel or pebbles can be used to form the layer, or specific industrialised components can be used for this drainage purpose.

- **Waterproofing layer:** Important for protecting the building's structure from infiltration, keeping the interior dry.

Special attention must be paid to the application of the eco-roof components, as the structural support used must withstand both permanent and accidental loads (ARAÚJO, 2007), i.e. consider, for example, the increase in load caused by water saturation in the structure when precipitation occurs, because the eco-roof has the characteristic of retaining rainwater, reducing surface runoff.

Another important factor to consider is the slope of the green roof, which is directly linked to its ability to retain water and slow down surface runoff. This slope varies according to the type of green roof used; the extensive type, which was used in this study, has a greater possibility of slope. Morais & Roriz (2005) state that extensive green roofs allow a slope of 0 to 30°, i.e. from 0 to approximately 16%.

There are several studies in the literature aimed at determining the thermal performance of eco-roofs. In his study, Beyer (2006) used wooden boxes to analyse the thermal behaviour of ecotiles and compared them with conventional roofs (figure 7). This study concluded that the use of ecotiles is responsible for significant thermal insulation of the roof, reducing the heating of the tile and the air beneath it.

Figure 7: Experiments to measure the performance of the Eco-roof.
Source: Beyer, 2006.

When applying the green roof, Andrade (2007) used the layers shown in figure 8. In his work, he studied the thermal behaviour of a single-storey building, where two types of grass were applied, Brachiaria grass *(Brachiaría humidicola)* and Emerald grass *(Zoysia japonica)* to check the differences between the temperatures of the slab.

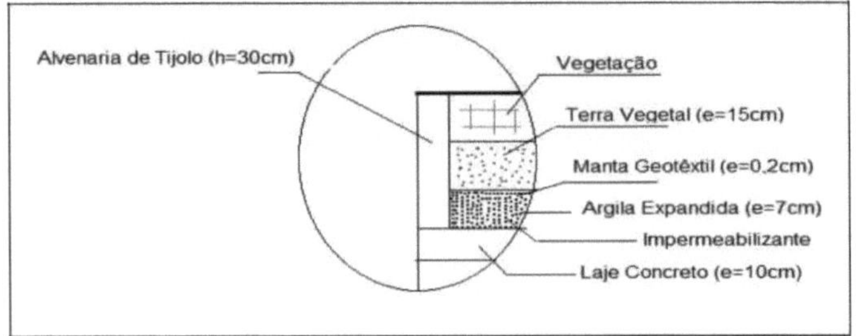

Figure 8: Component layers of the analysed green roof.
Source: adapted from Andrade, 2007.

In the research by Lopes (2007), an analysis of the thermal behaviour of a Light Green Roof (LGV) system was carried out, comparing it with different roofing systems in various test cells, as shown in the following figure.

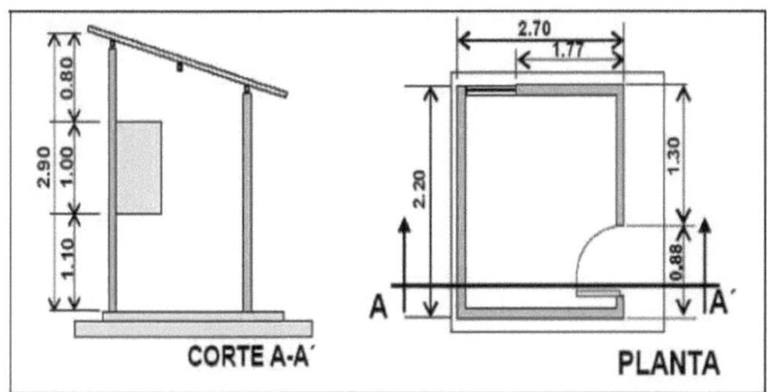

Figure 9: Schematic section and plan of the test cells.
Source: Vecchia & Castaneda, 2006 apud Lopes, 2007.

In the test cell, the green roof was installed, with a 10% slope, in which the following layers were found: Support structure, where the structure was implanted; Waterproof layer, made with a paste consisting of cement, water and latex, with the subsequent application of a vegetable polymer resin based on castor oil; Drainage layer, where a geocomposite called MacDrain 2L was used, manufactured by the company Maccaferri; and finally the substrate and vegetation, where emerald grass *(Zoysia japonica)* was used, planted in a substrate consisting of prepared soil. The complete system can be seen in the figure below.

Figure 10: CVL system in test cells at the USP - São Carlos experimental site.
Source: Lima, 2005 apud Lopes, 2007.

Green roofs are not only important for reducing the temperature inside homes, but they can also mitigate urban flooding problems, as presented in the work by Bacovis (2010), who aimed to assess the hydrological performance of a green roof in an urban environment in the process of mitigating flooding. To this end, a prototype green roof tile was built (the layers used in its design are shown in figure 11). In this prototype Bacovis (2010) uses 15 mm wood and a waterproofing tarpaulin to represent the waterproof roof tile, the drainage layer is represented by a layer of expanded clay between two layers of bidim RT 07 geotextile blankets, above this there is a layer of earth which is the substrate and the vegetation layer, in which São Carlos grass *(Axonopus compressus)* was used.

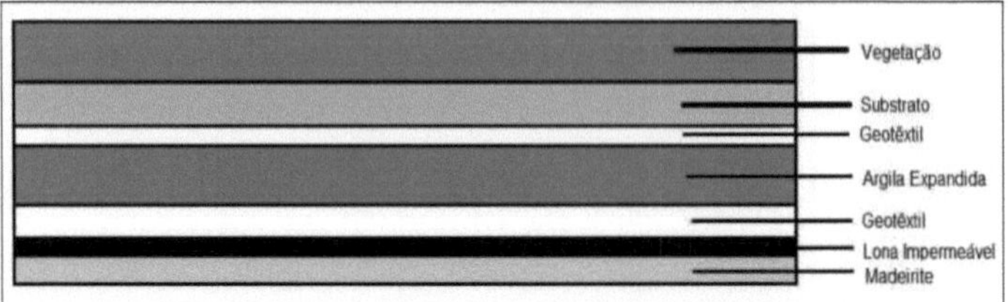

Figure 11: Layers of the extensive green roof prototype.
Source: adapted from Bacovis, 2010.

The research concluded that eco-roofs reduce a portion of the volume of surface runoff as well as increasing the detention time of rainwater, so they present themselves as a potential alternative to traditional models. Figure 12 shows the construction stages of the prototype

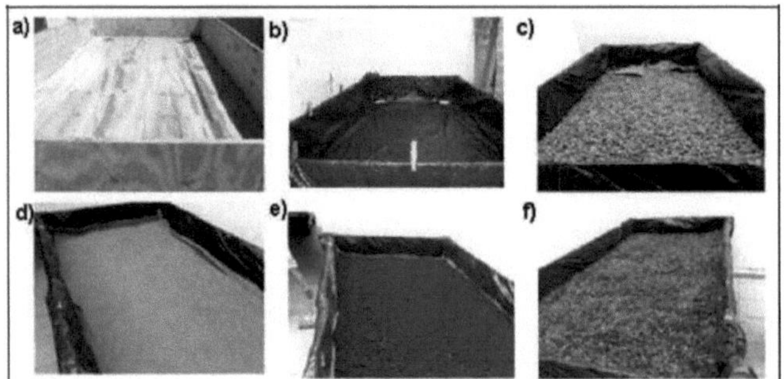

Figure 12: Layers of the extensive green roof prototype; a) Madeirite; b) Madeirite covered by the waterproof tarpaulin; c) Madeirite covered by the geotextile and expanded clay d) Second layer of geotextile; e) Substrate; f) Vegetation
Source: Bacovis, 2010.

Vacilikio & Fleischfresser (2011) made a comparison between green and conventional roofs in terms of internal room temperatures. In this study, two identical wooden boxes were built with a sloping fibre cement roof, and one of the boxes had its roof covered with grass. Internal temperature variations were measured every 15 minutes in these boxes using Minipa MT-600 digital thermometers. From the results obtained, they concluded that the eco-roof contributes to the thermal comfort of an indoor environment.

3.5- Benefits

Nowadays, large urban centres suffer from various problems related to urbanisation. Excessive heat, soil sealing, increased pollution and a reduction in green areas are some of the problems faced in these centres today. The use of green roofs,

replacing conventional roofing, appears to be a technique capable of alleviating several of these problems.

Heneine (2008) states that there is no doubt that green roofs have a wide range of substantial advantages over traditional roofs. Knowledge of these benefits is therefore of fundamental importance in the search for alternatives to solve the problems of urban centres.

In the case of ecotiles, the advantages of their use range from aesthetic and social factors to environmental and economic ones, and within the literature, the most studied benefits are the reduction of surface runoff, the mitigation of temperature variations and the increase in green area, in addition to the aesthetic and social effects.

3.5.1 - Temperature regulation

Green roofs are known for having a thermal behaviour that mitigates the effects of harsh climates, such as excess heat. This is due to the fact that this type of roof does not have a large temperature range, i.e. it tends not to have a large variation between the maximum and minimum temperatures, compared to other types of roof tiles (VECCHIA, 2005).

In equatorial climates, where solar radiation is very intense, high temperatures cause thermal discomfort. The use of eco-roofs combats this by reducing heat transmission into buildings. Krunche (1982 in ALMEIDA 2008) states that only 13 per cent of the radiation is transmitted to the support base, around 27 per cent is reflected, 60 per cent is absorbed by the plants and the substrate through evapotranspiration.When compared with other types of roofing, it can be seen that the variation in surface temperatures of eco-roofs is considerably less than with other types of roofing, as can be seen in figure 13.

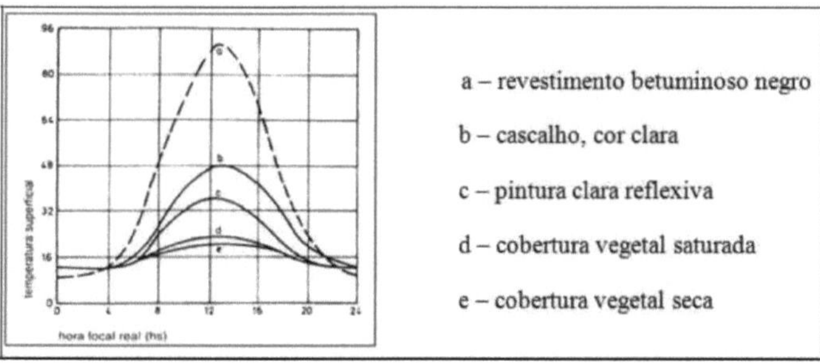

Figure 13: Temperatures (°C) measured over the course of a day on different surfaces above a flat roof on a sunny summer's day in a building in Germany.

Source: Gertis, 1977 apud Almeida, 2008.

Temperature regulation generates a thermal benefit both inside buildings and around green roofs, providing better thermal comfort for the population, thus improving quality of life. Within urban spaces as a whole, cooling the air outside buildings combats the creation of heat islands, a phenomenon characteristic of urban centres (GOULART, 1994).

In addition to this increase in thermal comfort, the insulating characteristic of the vegetation cover generates economic benefits. This is because less heating reduces the use of air conditioners, thus reducing energy costs. Studies show that, despite the high cost of installing eco-roofs in a building, in the long term it can become viable considering that their application would result in a reduction in energy costs previously used to cool the environment (MELLO, COSTA, ALBERTI & FREITAS FILHO, 2010).

Wong (2003) stated in his study that a reduction of 0.11 to 0.84°C could be achieved in Tokyo (Japan) by implementing green roofs on 50 per cent of buildings, which would generate energy savings of around 1.6 million per day.

3.5. 2- Reducing Surface Runoff

Another major problem in urban centres is the increase in surface runoff, which can lead to flooding, aggravated by places that don't have drainage infrastructure. This is because rainwater cannot infiltrate the soil due to waterproofing, so it flows through drainage systems into the rainwater networks. Heneine (2008) stipulates that approximately 75% of rainwater in city centres is directed directly into rainwater networks, compared to 5% in forested areas.

According to Schueler (2001 in Almeida, 2008), the modern concept of combating floods is tied to the need to bring the volume of run-off in urban basins as close as possible to the values prior to uncontrolled and disordered occupation and urbanisation (figure 14).

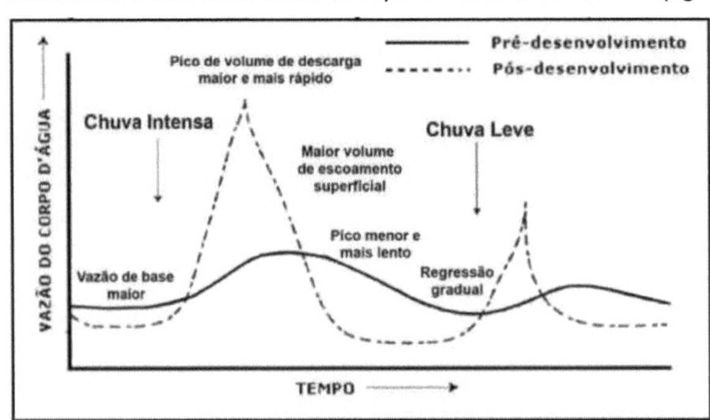

Figure 14: Effects of urbanisation on surface runoff.
Source: ambientebrasil, 2012.

Green roofs can improve this aspect because they have the capacity to reduce the amount of water that runs off the surface, as well as slowing down this runoff, preventing the rainwater galleries from receiving all the rainwater loads at once.

The decrease and delay in rainwater runoff by green roofs can be explained by the same reasons as the lower runoff found in forested areas: the existing vegetation uses this water for its vital functions and then eliminates it through evapotranspiration, reducing the water load, and the soil in green roofs retains and delays water from flowing directly into drainage systems (GOULART, 1994).

Figure 15 provides an easy-to-visualise example of the path taken by water on two types of roof, green and conventional:

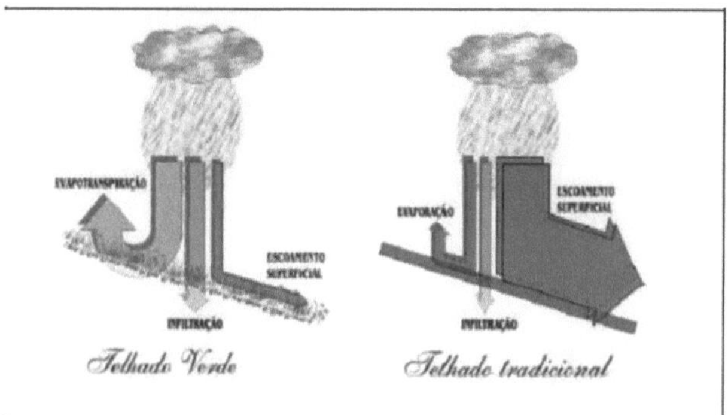

Figure 15: Comparison of runoff from a green roof and a normal roof.
Source: Laar, 2002.

3.5. 3- Increasing Green Areas

According to Araújo (2007), this issue is already of concern to large urban centres, which in turn are building more and more urban parks and encouraging the maintenance of green areas. The use of eco-roofs is an effective way of both maintaining and increasing green areas.

Almeida (2008) says that "The use of green roofs can act to integrate the building with its surroundings, and can act as a focus for the dissemination of fauna and flora, restoring, in a way, the conditions of the original biotopes". However, the vegetation to be used must be chosen correctly so as not to be affected by insects or birds that could transmit some kind of disease.

3.5.4- Aesthetic and social effects

CREA-RJ's magazine has an article on green roofs that lists several advantages of using them. In it, agronomist João Manuel Linck Feijó, from the Rio Grande do Sul company Ecotelhado, says that "This natural cover has several benefits over roof tiles. The vegetation acts as a thermal insulator, as it releases the heat absorbed through photosynthesis, providing a pleasant environment. In addition, this type of roof has an aesthetic advantage, which contrasts with the scenery of large cities, where green landscapes are increasingly rare" (REVISTA CREA-RJ, 2008). The aesthetic and social effects he refers to are one of the advantages associated with the use of green roofs. Architectural projects are increasingly seeking harmony between built environments and the environment, and the use of green roofs is one method of achieving this goal. Urban roofs have the potential to achieve this balance by becoming lawns, gardens (figure 16) or horticultures and harbouring a wide variety of plant species with adequate irrigation, and even where roofs are inaccessible but visible, plantations are beneficial due to the therapeutic effect that the presence of green plants and nature bring (HENEINE, 2008).

Figure 16: Roof of a building in Portland, a city in the north-west of the United States.
Source: Greenroofs, 2012.

In addition, these effects also extend to poor communities, where they can become an interesting alternative for the process of citizenship and commitment to the environment on the part of populations that lack environmental education and sanitation, presenting a potential for improvements in the infrastructure of these communities (MARY, 2008).

3.6- Main Barriers to Use

According to Happe (2005), the main barriers to the construction of natural roofs are the lack of professionals with the necessary technical skills, the cost of the materials that make up the eco-roof and the specific labour required for its installation and periodic maintenance.

Focusing on the issue of cost, Porsche and Kõhler (2003) concluded in their study of the relationship between the costs and benefits of green roofs that the cost depends on the materials used, which have an influence on their durability.

According to Wong et al. (2003), although green roofs may cost more initially, this amount is later offset by the benefits derived from the useful life cycle of this roofing system.

According to Peck et al. (1999), all new technologies face certain barriers in their introduction to the market. For the author, these difficulties include a lack of knowledge and awareness of the subject, a lack of incentives for implementation, cost-benefit barriers, as well as the technical risks associated with the uncertainty of the production system.

It can therefore be concluded that most of the benefits of green roofs are measured over the long term, making it difficult to determine costs for a quick financial assessment.

3.6.1- Maintenance and care

Another barrier to the spread of green roofs is the cost and labour involved in maintenance and care. Virtually all types of green roof require some kind of maintenance. The frequency of this maintenance varies and depends on the type of roof used. In general, green roofs with a low self-weight need little maintenance and irrigation once their vegetation is fully established (LOPES, 2007). However, in general, the frequency of

Irrigation is directly linked to the depth of the soil; shallow soils require repeated watering during periods of drought and intense summer (NASCIMENTO, 2008).

In her study, HENEINE (2008) states that in order to guarantee a long period of functionality and aesthetics of the green roof, it is necessary to provide maintenance measures. The author divides them into three stages:

- Installation maintenance: this is the first care taken with the installation of the vegetation. Plenty of water is needed for the dry seasons and care must be taken to prevent the proliferation of weeds.

  Maintenance development: the same maintenance as the installation, but with less intensity.

  Keep up the maintenance: once the vegetation is completely in place, it is necessary to carry out maintenance once or twice a year to, among other things, trim the perimeter of the cover and remove weeds.

CHAPTER 4

THERMAL COMFORT

People need to feel comfortable in a particular environment, and this feeling of well-being is defined as environmental comfort. The concept of environmental comfort is relative, broad and subjective, as each person reacts differently in a given environment. According to PICADA (1999:2) "Environmental comfort is inextricably linked to the psychosomatic needs of the individual, which often have to be expressed in order to be met and, at other times, because they are so specific and particular, are relegated to the generic solution adopted."

Since the concept of environmental comfort is very broad, several parameters make up its concept, such as visual, acoustic, sound, light and thermal comfort. Table 3 shows examples of the requirements of some of these parameters according to specific standards.

Table 3: Examples of environmental comfort requirements

Requirements	Criteria for good performance or fulfilment of a mandatory requirement	Document of Reference
Thermal comfort	Maximum percentage of people dissatisfied with air movement Maximum horizontal radiant temperature asymmetry < 10°C	ISO 7730
	Maximum internal operating temperature as a function of the average monthly external temperature, for non-air-conditioned environments	ASHRAE 55/2005
Comfort Acoustic	Maximum noise level inside bedrooms of 45dB(A)	NBR 140152
	Maximum reverberation time in classrooms between 0.6s and 0.7s, depending on the volume of the room	ANSI-ASA S12.60
	A-weighted total sound pressure level greater than the Speech Interference Level	ISO 9921-1
Comfort Visual	Minimum illuminance level of 500 lux in office building workspaces	NBR 5413
	Possibility of having visual contact with the outside world	DIN 5034-1
	maximum glare index of 19 caused by the artificial lighting system in waiting rooms.	AS/NZS 1680.2

Source: adapted from Techne Magazine, 2007.

The ASHRAE 55 standard states that: "Thermal comfort is that mental condition which expresses satisfaction with the thermal environment." The condition of thermal comfort that a person is subjected to in an indoor environment is standardised by the International Standard ISO 7730 (1994) (XAVIER, 2000).

Thermal comfort is directly linked to heat exchange between the human body and the environment. Humans feel better when they lose the heat produced by their metabolism to the environment, without resorting to any thermoregulation mechanism (FROTA E SCHIFFER, 2003).

Thermal comfort can also be analysed according to the environmental aspect, which is the combination of physical variables inherent in the environment that create thermal environmental conditions so that the fewest number of people are dissatisfied with the environment. Therefore, an environment is said to be thermally acceptable when it presents combinations of physical variables that make it uncomfortable for the fewest number of people possible (ISO 7730, 1994 apud XAVIER, 2000).

Some climatic variables are related to thermal comfort, such as temperature, humidity, air speed and incident solar radiation. These are closely related to rainfall patterns, vegetation, soil permeability, surface and underground water, topography, among other local aspects susceptible to alteration by human presence (FROTA E SCHIFFER, 2003).

According to Auliciems and Szokolay (1997), Socrates, in the 4th century BC, already had some ideas about the climatic suitability of homes and how to build to ensure thermal comfort, and Vitruvius, in the 1st century AD, considered it necessary to consider the climate in building projects for reasons of health and comfort (XAVIER, 2000).

In England, Vemnon and Warner (1932) and later Bedford (1936) carried out empirical studies involving factory workers to establish the limits of environmental conditions for work. Olgyay (1963) was the forerunner in bringing together the results of various areas of study in an attempt to define a thermal comfort zone (XAVIER, 2000).

In 1916, the American Ventilation Commission carried out studies that sought to relate and determine the influence of temperature and humidity conditions on work performance. They found that for physical labour, increasing the ambient temperature from 20°C to 24°C reduces performance by 15% and when the ambient temperature is 30°C, with 80% relative humidity, performance drops by 28% (FROTA & SCHIFFER, 2003).

Over the years, architecture has evolved to favour the needs of the individual. Form and function are no longer the only objectives of a building. Now energy efficiency and environmental requirements must also be taken into account in developments that aim to achieve high levels of satisfaction (NERBAS, 2009).

Architecture and urban planning must always consider climatic and environmental factors when designing homes, buildings or cities in order to provide comfort in the living space, as well as thermal, lighting, acoustic, water and soil quality comfort. When thermal comfort refers to architecture, it is linked to the basic issue of providing built human environments with the necessary conditions for habitability, using available resources rationally (RIBEIRO, 2002). From this line of thinking comes sustainable construction or green construction.

CHAPTER 5

METHODOLOGY

The methodology of this work was defined in the construction of the measurement units, the construction of the eco-roof layers and the collection of data. These steps will be explained in detail in the following sections.

5.1 - Characterisation of the study area

As previously mentioned, this study aims to compare the behaviour of a conventional roof and an eco-roof in the equatorial climate of the northern region, analysing the thermal behaviour of the two types of roof and their influence on the internal environment and, based on the results, proposing the use of eco-roofs in Belém.

Before starting the methodological process, it is important to add some climatological information about the Amazon region, where the work is located, and to specify the exact location of the measurement units.

The Amazon has an equatorial climate, hot and humid, with practically no dry season. Due to the high levels of incident solar energy, the air temperature shows little variation throughout the year, which means that the Amazon has an annual temperature regularity. Annual averages show temperatures of around 26 - 28°C (FERREIRA, SILVA DIAS & JUSTI DA SILVA, 2000).

The lowest average air temperatures occur in February, a month typical of the rainy season with high cloud cover. The highest average air temperatures occurred in October, which represents the transition between the end of the less rainy period and the beginning of the more rainy period (INMET, 1992).

In the tropical region, rainfall is the most variable meteorological element and the main factor used to subdivide climates. The rainfall regime in the state of Pará is well defined into a rainy season, from December to May, regionally known as winter, and a less rainy season, from June to November, regionally known as summer (OLIVEIRA, 2002).

Relative humidity is high, with values above 80 per cent in every month of the year. The highest relative humidity values occur in the rainiest quarter, at around 89%, as a result of the lower temperatures during this period. The average duration of hours of sunshine shows a well-defined seasonal variation, with maximum values in July (less cloudiness) and minimum values in February (more cloudiness). The average duration of sunshine varied from 3.8 hours in February to 7.5 hours in July (INMET, 1992).

Once built, the metering units were installed in a private condominium located in the Mangueirão neighbourhood in the city of Belém. This location was chosen because it met all the needs of the study: a vast plot of land without any type of construction where the metering modules could be installed, as well as the safety of the site in terms of protecting the units, the equipment and the authors themselves.

The condominium is on the banks of Centenário Street, very close to the Estádio Olímpico Estadual Jornalista Edgar Augusto Proença, known as Mangueirão. On one side is the Benguí neighbourhood and on the other the Parque Bosque Médice II (the green area on the map). Its geographical coordinate is 1°23'06.8"S 48°27'06.5"W. By transferring this to Google Maps, we get the image shown in the figure below.

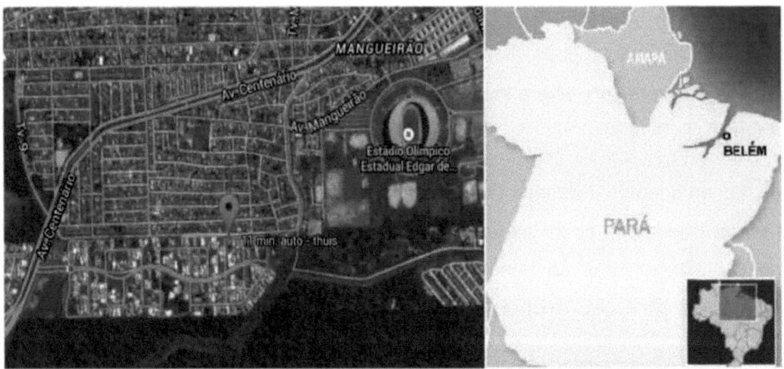

Figure 17: Location of the gated community where the metering units were installed.
Source: Google Maps, 2012.

Data collection lasted a total of 12 months - starting in December 2011 and ending in December 2012 - as the aim was to cover both the rainy season and the less rainy season, thus closing a complete meteorological cycle.At the same time as data was acquired at the measurement units, meteorological data was also acquired for the city of Belém, as explained in the next topic.

5.2 - Meteorological data collection

Vecchia (2003) states that in order to assess the thermal behaviour of a building's envelope, it is necessary to analyse the meteorological data obtained on the surface as a function of the time of exposure to climatic conditions. According to the author, the most important parameters are: air temperature and relative humidity, prevailing wind direction and speed, rainfall and solar radiation.

According to Lopes (2007), solar radiation is the main element in heat exchange processes, especially in the roofing systems of buildings.

Based on the teachings of Lopes (2007) and Vecchia (2003), this study analysed data on insolation, relative humidity, rainfall and air temperature in Belém throughout the data collection period. The data for these two parameters was obtained from INMET (National Meteorological Institute).

The institute has a modern automatic surface weather station, called BELEM-A201, consisting of a central memory unit ("data logger") connected to various meteorological parameter sensors (atmospheric pressure, sunshine, temperature and relative humidity, precipitation, solar radiation, wind direction and speed, etc.), which integrates the values observed minute by minute and makes them available automatically every hour. It is located inside INMET on Avenida Almirante Barroso (Latitude: -1.411228° Longitude: - 48.439512°). It has been in operation for a relatively short time, having opened at the beginning of 2013, and for information purposes, its WMO code is 81680. The figure below shows an example of an automatic weather station like the one in Belém.

Figure 18: Example of an automatic weather station.
Source: INMET (2014).

5.3 - Construction of Measurement Units

The measurement units, called UTC (Conventional Roof Unit) and UET (Eco-Roof Unit), were built based on studies by Beyer (2006) and Vacilikio & Fleischfresser (2011), and each had an internal volume of $1m^3$. The units were built from hardwood and their structure was screwed together so that they could be easily dismantled and assembled elsewhere, with the aim of reusing materials. The construction details of the measuring units can be seen in figure 19, which specifies their dimensions.

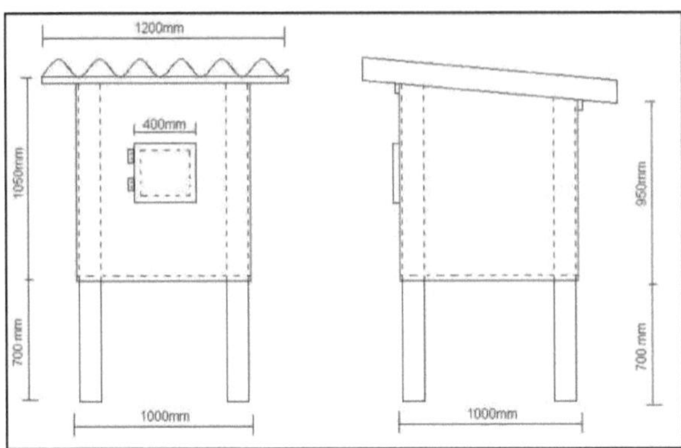

Figure 19: Schematic plan of the units.
Source: The authors, 2011.

You can see that a hatch has been built in front of the units to access their interior. A small shelf was placed inside to support the instruments and facilitate measurements. The only feature that differentiates these units is the roofing system: in one the eco-roof was installed and in the other the fibre cement tile.

Firstly, the structure of the units was assembled (figure 20). These structures have a height of 1.75 metres at the front of the units and 1.65 metres at the back. This difference was proposed in order to obtain a roof pitch in line with the literature reviewed. The slope of the roof tiles in this study was 5.7°, around 10%, the same slope used by Lopes (2007) in his research. This height was also aimed at raising the unit above ground level, as well as fixing the unit to the ground

Figure 20: Front and side view of the measuring unit.
Source: The authors, 2011.

The width and length were made one metre each to make it easier to obtain an internal volume of one cubic metre, where the thermometers would be installed and measurements taken.

Once the structure was ready, the 10mm plywood sheets were nailed down to form the walls and floor of the unit (figure 21). The front wall of the unit was the only one that differed from the others, as it had a door that gave access to the inside of the metering units. This door is 0.4 metres wide and 0.4 metres high, and was sized to allow the metering equipment to be installed and handled inside the units. A lock was also added to this door, allowing the units to be locked with a padlock, thus increasing the security of the equipment inside the units.

Figure 21: Fixing the walls to the structure.
Source: The authors, 2011.

The fibre cement tiles were then screwed to the structure. To finalise, the two units were painted white to reduce heat absorption from the walls of the units, and also for aesthetic reasons (Figure 22). A 15 cm shelf was made on the inside of the units, where the thermometers would be positioned.

Figure 22: Finalised unit .

Source: The authors, 2011.

5.4- Ecotile Module

The eco-tile module was assembled directly on top of one of the measurement units. The components of the structure follow what has been reviewed in the literature and are shown in figure 23.

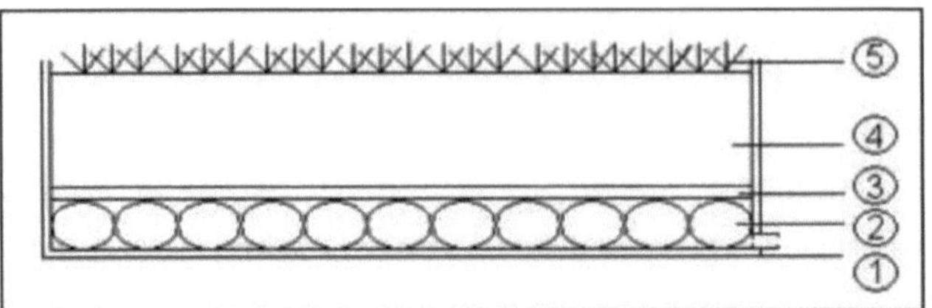

Figure 23: Components of the ecotile module: 1- Waterproof layer; 2- Drainage layer; 3- Filter layer; 4- Substrate; 5- Vegetation.

Source: The authors, 2011.

This construction was based on all the structures needed for the green roof to function properly. The structures considered and built are described below.

5.4.1- Waterproof layer

The waterproof layer in this study was represented by a fibreglass box 1.2m wide, 1.2m long and 0.2m high. Due to the need for a sturdy, waterproof structure to support the other layers of the module, a fibreglass box was used. In order to prevent water from accumulating inside the module, holes were drilled in the side of the box facing the lowest level (as shown in figure 24), so that the water that accumulated at the bottom of the module could drain away and not harm the existing vegetation.

Figure 24: Water drainage holes in the fibre cement box.
Source: The authors, 2011.

When installing the module, a support had to be installed on the back of the measuring unit so that the module could be held in place (Figure 25).

Figure 25: Supports on the back of the measuring units.
Source: The authors, 2011.

5.4. 2- Drainage layer

In this study, the drainage layer was represented by pebbles (figure 26), medium-sized stones that are generally used to decorate gardens. Smaller stones can also be used, but the smaller they are, the more difficult drainage becomes, due to the reduced space for water to flow.

Figure 26: Assembly of the drainage layer.
Source: The authors, 2011.

5.4. 3- Filter layer

The filter layer was represented by a geotextile blanket (figure 27), generally used in gardens for the same purpose. This layer lies between the drainage layer (pebbles) and the substrate layer, preventing soil from spreading between the voids in the pebbles, which would make it difficult for the water to drain away.

Figure 27: Addition of the geotextile blanket.
Source: The authors, 2012.

5.4.4- Substrate

The substrate layer was represented by black soil (figure 28), ideal for gardening. The soil was ideal for this study as it had ideal characteristics for planting vegetation, without the need to add nutrients or other types of soil correction.

Figure 28: Addition of black earth.
Source: The authors, 2012.

5.4.5- Vegetation

The type of vegetation used in this study was emerald grass *(Zoysia japonica).* The vegetation was the last component to be placed, finalising the assembly of the ecotile module, as can be seen in the figure below.

Figure 29: Adding vegetation.
Source: The authors, 2012.

Emerald grass originated in Japan and is widely adapted to Brazilian conditions, and can be used from north to south. It is resistant to trampling and adapts to different types of soil. It also has a strong root system and rhizomes (GURGEL, 2003), and because it has these characteristics, it was chosen to be used in this study.Once the ecotile module had been assembled, the stage of building the measuring units was over (figure 30 shows the final result), and it was then possible to start the data acquisition stage.

Figure 30: Measuring units ready and installed in the field.
Source: The authors, 2011.

5.5- Measuring equipment

The thermometers were purchased according to the functionalities required to carry out the measurements. We tried to acquire a *datalogger* thermometer with similar features to the Minipa MT-600 model used by Vacilikio & Fleischfresser (2011) in their study, but the same model could not be purchased as it was obsolete and no longer manufactured. The equivalent thermometer found on the market was the Icei TD-880 (figure 31). It had all the features of the thermometer used by Vacilikio & Fleischfresser (2011), as well as good battery life and a system for organising the data acquired.

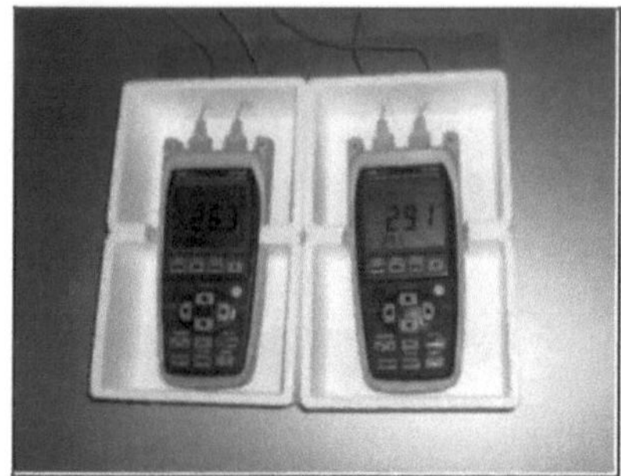

Figure 31: Photo of the thermometers.
Source: The authors, 2012.

The TD-880 thermometer has dimensions of 164x76x32mm and a weight of 415g (including battery), its display is a four-digit LCD display and it shows the temperature in Celsius, Fahrenheit and Kelvin. It measures on two channels simultaneously and is compatible with various types of thermocouple (K, J, T, E, R, S and N). Its *datalogger* has a storage capacity of 10,000 data points. Its autonomy is 550 hours with alkaline batteries and it is powered by four 1.5 V AAA batteries (Icei, 2012).

Considering these characteristics, the Icei TD-880 thermometer was able to satisfactorily fulfil the needs required in this study. Measurement on two channels simultaneously was essential for acquiring data on both the internal air temperature and the internal surface temperature of the tiles in the measurement units. The *datalogger* has a high data retention capacity, which meant that a shorter time between measurements could be adopted without running the risk of reaching maximum capacity and causing data loss or other problems. The battery life ensured the thermometer had plenty of autonomy, reducing the need to remove the thermometer from the measuring units when transferring the thermometer's data to the computer.

In addition, the thermometer had functions for 'freezing' the reading, relative measurement, recording the minimum, maximum and average, difference between channels, counter (up to 99 hours and 59 minutes), alarm for high and low values and automatic switch-off.

Data is transferred to the computer via a cable with an RS-232 interface (shown in figure 3), which comes with the product, along with a CD for installing the data transfer programme, Temp monitor (figure 31). Using Temp monitor, it is possible to save the data in Excel spreadsheets, which, by obtaining the data acquired in a given time interval by the two channels, highlights the maximum and minimum temperature of each.

Figure 32: RS-232 cable.
Source: The authors, 2012.

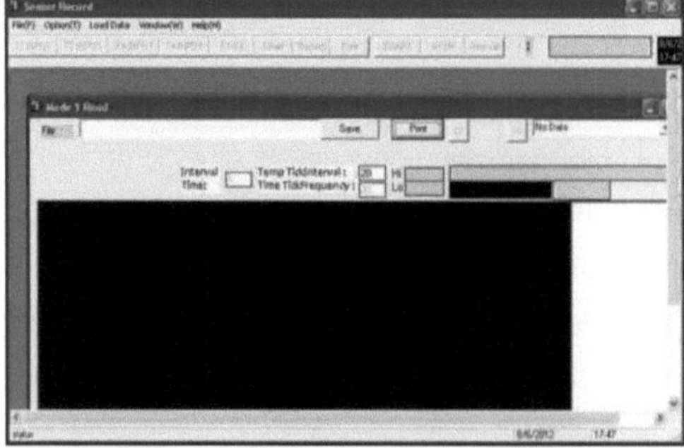

Figure 33: Temp monitor programme.

Source: The authors, 2012.

5.5.1- Equipment installation and data acquisition

Inside the measuring units, space was dedicated to installing the equipment. A shelf was built in the middle of the internal space, making it easier to handle the equipment (figure 34).

Figure 34: Internal shelving unit.
Source: The authors, 2012.

To protect the equipment (against temperature variations and other factors), it was placed inside a Styrofoam box (figure 35). Two holes were drilled in the Styrofoam box to allow the thermocouples to exit.

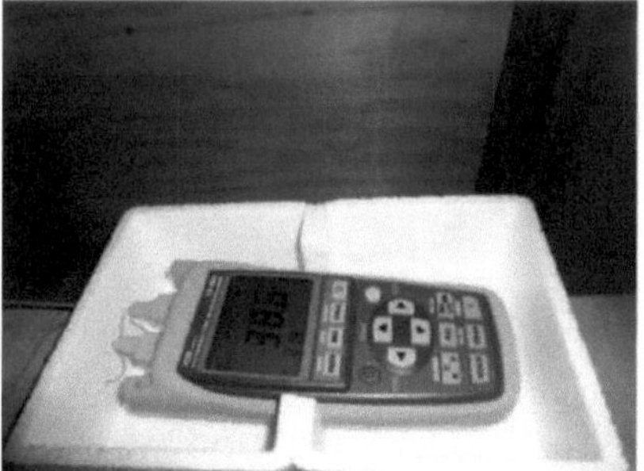

Figure 35: Thermometer in the protective styrofoam box.
Source: The authors, 2012.

In each measurement unit, two types of data were collected simultaneously: the internal temperature of the air and the surface temperature of the tiles. In this way, the tip of a thermocouple was positioned in direct contact with the inner surface of the tile, and the other tip was positioned in the middle of the box, in contact only with the air, in order to measure the temperature of the internal air (figure 36).

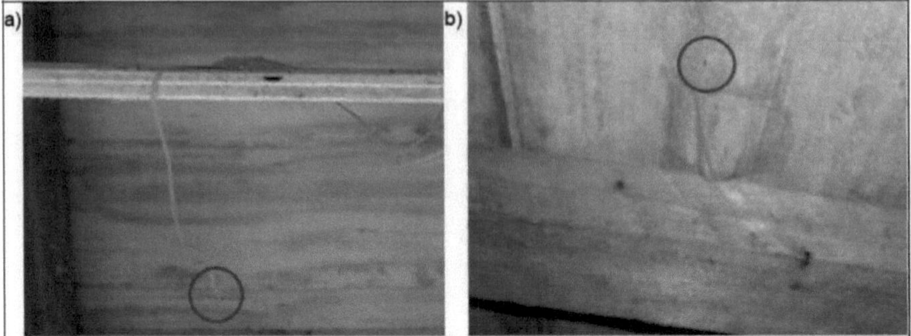

Figure 36: Measuring points inside the units: a) Internal air and b) Internal surface of the tile.
Source: The authors, 2012.

In order to define in detail the behaviour of the tiles during the measurement time, the data was acquired uninterruptedly (except when changing batteries and transferring data), with a 10-minute interval between them. This means that every hour, six data points were obtained regarding the temperature variations that occurred in each of the channels of the measuring equipment.

As the battery life allowed the thermometers to last a long time, the data was transferred every fortnight.

At the same time, meteorological data was acquired from the city of Belém (as mentioned above) and organised into tables so that it could later be compared to the data on thermal behaviour in the measurement units. Due to the large amount of data, it was analysed according to its daily average, maximum and minimum. In this daily analysis, a 3-day period was chosen in which the temperature reached more significant levels, in order to analyse the variations obtained every 10 minutes, highlighting the behavioural differences of the different types of roof tiles analysed.

CHAPTER 6

RESULTS AND DISCUSSION

This chapter analyses and interprets the data acquired. Considering the long period of data acquisition, which provided information on an entire meteorological cycle, a valuable amount of data was obtained - approximately 105,000 - and in order to better visualise all this information, this data was organised in the form of graphs (Annex 1 shows the average monthly data obtained over the entire measurement period). It should be noted that in addition to the 105,000 data points resulting from the measurement units, there is also meteorological data that has been analysed in conjunction with them. At the end of July and the beginning of July, totalling two weeks, there were some data losses due to problems with the equipment, but the study was not jeopardised.

Due to the wide range of possible analyses, the average, maximum and minimum daily temperatures of both the internal temperature of the measuring units and the internal surfaces of the roof tiles were analysed. The daily variation is a very detailed way of analysing the thermal behaviour of the measurement units, noting the interference of periods of high rainfall, radiation peaks and other meteorological variations.

Another parameter analysed during this study was the relationship between meteorological factors - outside air temperature, sunshine, relative humidity and rainfall - and the temperatures obtained in the measurement units. The aim was to relate the effects of these climatic parameters on the temperatures obtained.

In addition, a 3-day period was defined in which the highest temperature levels were obtained, making it possible to analyse in more detail and significantly the behaviours that occur over a 24-hour period. This detailed study analysed all the variations that occurred in the measurement units every 10 minutes, studying the relationship with meteorological factors.

6.1- Benefits of the Eco-Roof Compared to the Conventional Roof

Looking at the average indoor air temperature graph below, you can see that there is a predominance of higher temperatures in the Conventional Roof Unit (CTU) when compared to the temperatures in the Eco-Tile Unit (ETU), with average temperatures of around 27.6° C and 27.4° C respectively. However, this difference is not so significant, around 0.26° C on average, but it is relevant if we take into account the significant period of time, as was the case in this research.

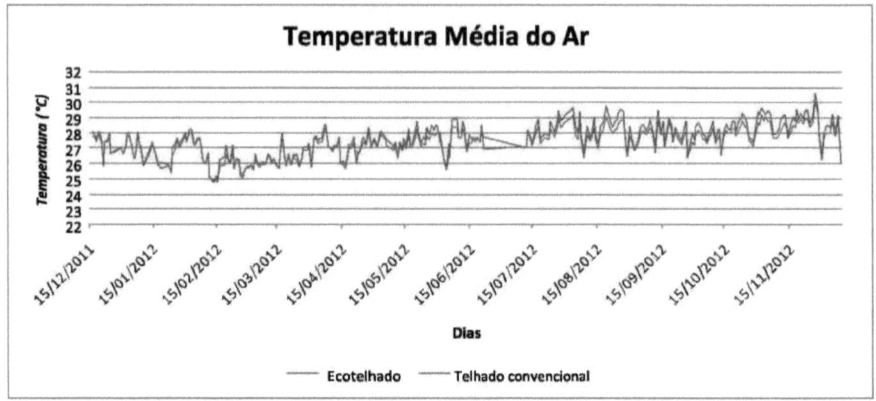

Graph 1: Average air temperatures in both units from December 2011 to December 2012.

Source: The authors, 2014.

Graph 1 also shows that there were no significant variations in temperature when comparing the readings obtained inside the Experimental Units (UCT and UET). This is possibly due to the thinness of the walls of the measurement units, which were made of 10mm plywood, meaning that there was still significant interference from the external environment inside the units.

Still analysing the average daily behaviour, it can be seen in the graph of the average internal surface temperature of the roof tiles (graph 2) that there is a greater difference in temperature when comparing the two units. The conventional roof unit shows much higher values (28.9°C), with an average of 1.5°C more than the temperature measured in the eco-roof unit (27.45°C).

This significant temperature difference demonstrates the eco-roof's characteristics as a temperature attenuator.

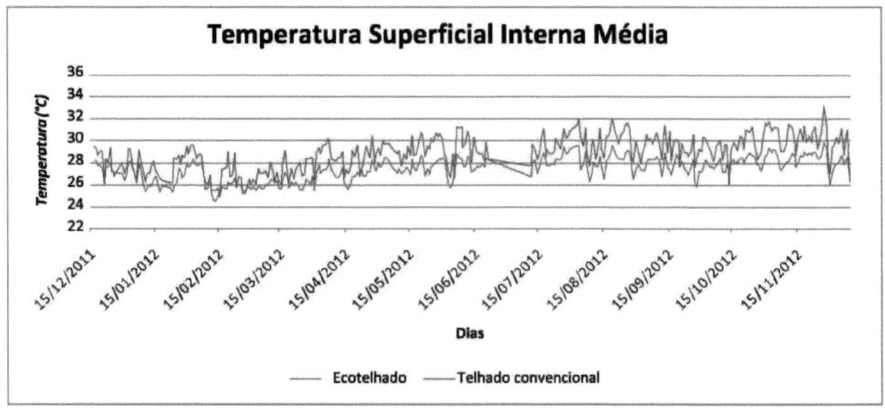

Graph 2: Average temperatures of the internal surfaces of the roof tiles at UTC and UET from December 2011 to December 2012.

Source: The authors, 2014.

It can also be seen that in the first few months of measurement - December to June - the temperature differences are smaller and more inconstant, while in the following months - August to December - greater temperature differences are achieved and for a longer period.

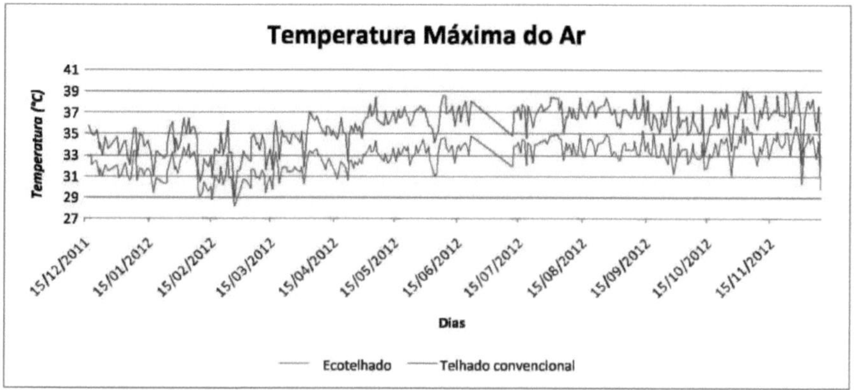

Graph 3: Maximum air temperatures at UTC and UET throughout the measurement period

Source: The authors, 2014.

When you analyse the maximum air temperatures (graph 3), you can see a big difference between the measurement units. The UTC has approximately 3°C higher maximum indoor air temperatures than the ETU. This behaviour is due to the heat absorption and dispersion characteristics of the ecotunnels, preventing it from reaching very high temperatures at times of peak solar radiation (between 11 am and 2 pm).

The biggest temperature difference found during the study period was 5.2 °C in September, when the air temperature was 37.4 °C at UTC and 32.2 °C at UET. It is important to emphasise that this was a period of low relative humidity, and the day in question was the driest of the entire period studied, with relative humidity of 54%. A more detailed analysis of the relationship between the temperature measured in the measurement units and the relative humidity will be covered in section 5.2 of this paper.

By analysing the maximum temperatures of the internal surfaces of the roof tiles (graph 4), it is easier to see how significant the difference in heat absorbed by the different types of roof tiles is.

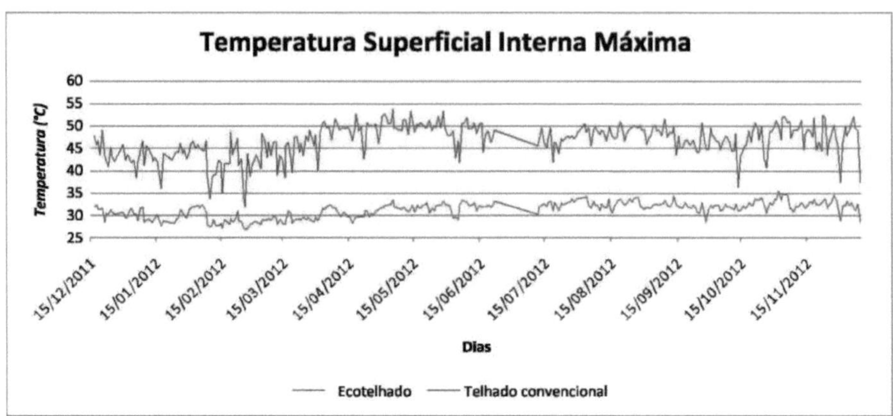

Graph 4: Maximum temperatures of the internal surfaces of the tiles at UTC and UET throughout the measurement period

Source: The authors, 2014

The average difference between the maximum temperatures on the inside surfaces of the roof tiles is 15.5°C, with UTC reaching peaks of 53.9°C, while at UET the

The maximum temperature reached on the tile surface was 35.4 °C. The greatest differences in temperature can also be seen in April and May, with a peak of 23.2°C in April.

When analysing the minimum air temperatures (graph 5), it can be seen that there is a small difference between the minimum air temperatures between the units. However, UTC has the lowest temperatures. This is due to the characteristics of the fibre cement roof tile, which not only heats up faster, but also loses heat more quickly, reaching lower values at night. This characteristic was another reason why there were no major differences in the average internal air temperatures of the units.

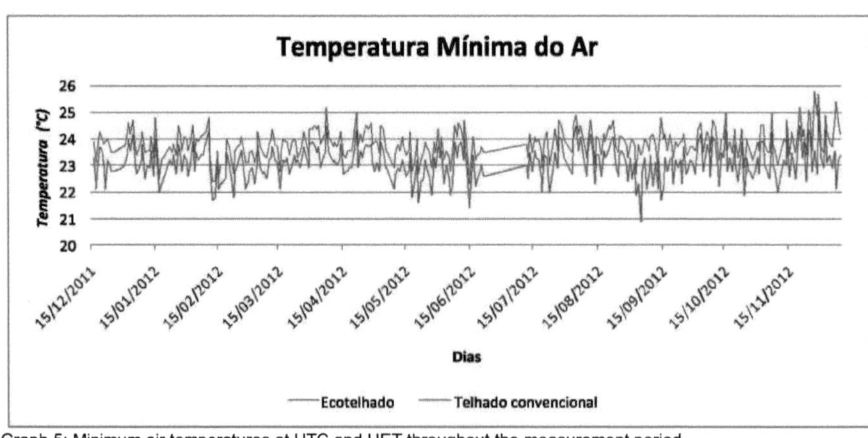

Graph 5: Minimum air temperatures at UTC and UET throughout the measurement period.

Source: The authors, 2014.

The average difference between the minimum internal air temperatures of the roof tiles is 0.76 °C, with the lowest temperature found during the study period being 20.9 °C at UTC and 22.3 °C at UET.

Analysing the minimum temperatures of the internal surfaces of the roof tiles (graph 6) makes it easier to understand the thermal characteristics of ecotiles, which also vary more slowly in terms of heat loss.

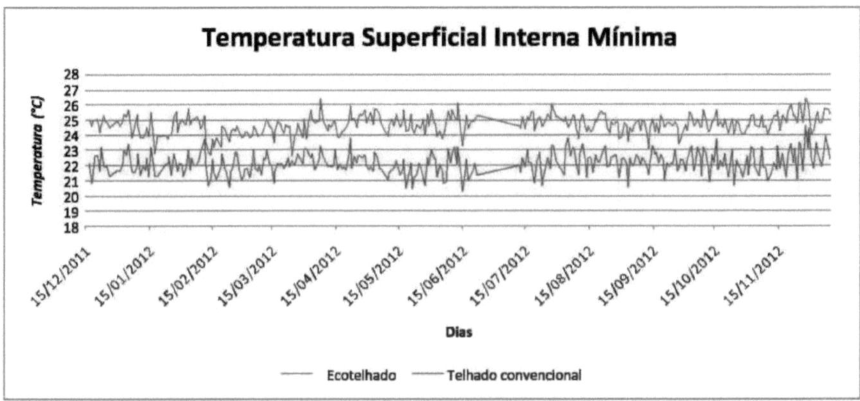

Graph 6: Minimum temperatures of the internal surfaces of the tiles at UTC and UET throughout the measurement period.

Source: The authors, 2014.

The average difference between the minimum temperatures on the inside surfaces of the roof tiles is 2.5°C, with UTC reaching minimums of 20.3°C, while at UET the minimum temperature reached on the roof tile surface was 22.6°C. It can also be seen that the variation in minimum surface temperature between the tiles is not as great as the maximum surface temperature. It is believed that this is because very low temperatures are not reached in the region.

Once this analysis has been carried out, it can be seen that the use of ecotiles makes it possible to achieve lower air and internal surface temperatures during the day than the use of conventional fibre cement roofing [table 4]. However, fibre cement roofs achieve lower temperatures at night than eco-roofs.

Table 4: Maximum, average and minimum air temperatures of the internal surfaces of the measurement units.

	Air Temperature (°C)			Internal Surface Temperature (°C)		
Tile type	Average	Maximum	Minimum	Average	Maximum	Minimum
Ecotile (UET)	27,4	36,4	22,3	27,5	35,4	22,6
Conventional roof (UTC)	27,6	39,1	20,9	28,9	53,9	20,3

Source: The authors, 2014.

6. 2- Evaluation of Meteorological and Temperature Parameters

This study collected meteorological data for the entire study period (Appendix 2) in order to analyse the influence of different parameters on the surface temperature of the measurement units. This revealed a significant relationship with the following parameters: outside air temperature, humidity, solar insolation and rainfall. These parameters have a connection with each other, but they were all analysed individually in order to have a more complete analysis of the influence of each one on the data obtained in this study.

Graph 7 below shows the relationship between solar insolation and the average internal surface temperature of the measuring units. It can be seen that in the first few months (up to May) the insolation time is shorter, averaging 5.65 hours, while in the following months the average rises to 7.9 hours. This was also reflected in the internal surface temperature of the measurement units, with the ETU showing an average of 26.8 °C until May and 28 °C in the following months, and the CTU showing 28 °C until May and 29.8 °C in the following months.

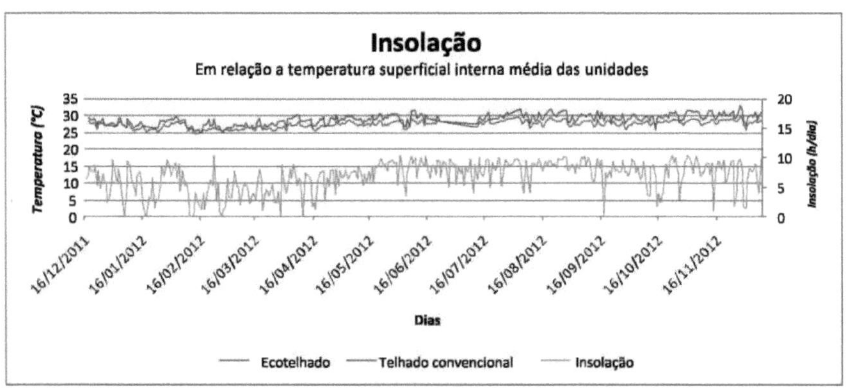

Graph 7: Relationship between solar insolation and the average internal surface temperature of the measurement units

Source: The authors, 2014.

It can therefore be seen that the longer the different roofs are exposed to solar insolation, the higher their temperature and the greater the heat passed into the units. It can also be seen that UTC is more sensitive to this parameter, since the temperature differences found are greater.

significant in the conventional cover unit, when comparing the period of greatest insolation with the period of least.

Graph 8 below shows the relationship between air humidity and the average internal surface temperature of the measuring units. It can be seen that, in general, the higher the humidity, the lower the temperatures on the roofs, i.e. in the period up to May, where the humidity is higher (average of 85.65%), the ETU and CTU had averages of 26.8 °C and 28°C respectively, and in the following months, with lower humidity (average of 77.9%), the ETU and CTU had averages of 28 °C and 29.8 °C respectively.

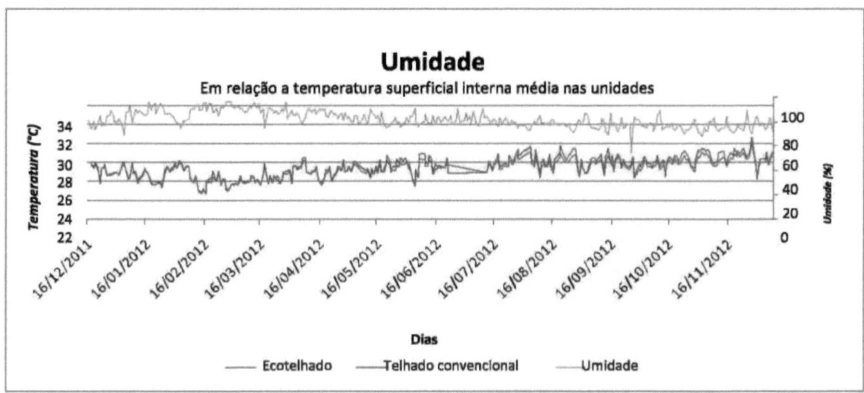

Graph 8: Relation of humidity to the average internal surface temperature of the measuring units

Source: The authors, 2014.

Graph 9 below shows the relationship between precipitation and the average internal surface temperature of the measurement units. Precipitation was the parameter that showed the least relationship with the temperature variations of the units, which is due to the fact that it is a one-off, momentary event.

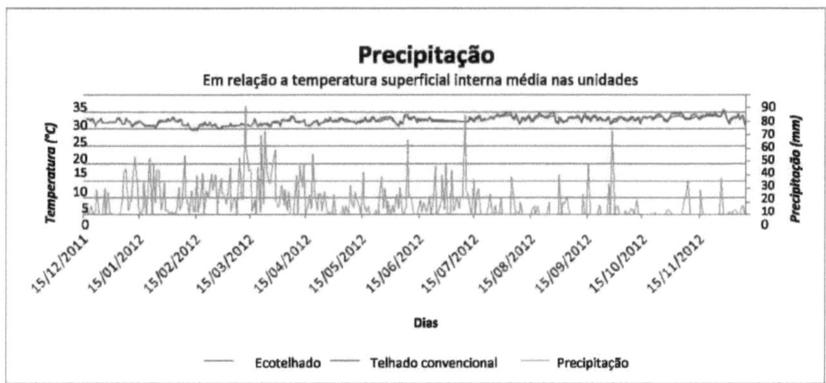

Graph 9: Relationship between rainfall and the average internal surface temperature of the measurement units

Source: The authors, 2014.

However, if we take into account the periods in which there is more or less rainfall, we can say that the rainiest periods are those with the lowest temperature values in the measurement units.

It is noticeable that towards the end of the measurements there was a decline in rainfall (specifically after 15/10/2012). This period of low rainfall was directly related to the maintenance of the eco-roof. As can be seen in figure 37, the grass died due to a lack of water. This occurred between 27 October and 10 November. As soon as the incident was realised, the vegetation was duly replaced. However, by analysing the temperatures recorded during the time the roof was dry, it was possible to see that they did not differ significantly from the temperatures recorded during the other periods. This is due to the fact that the dry roof still acts as a barrier to heat transmission, as well as helping to retain water that dissipates heat in the form of transpiration, so this data was kept and considered in the study.

Figure 37: UET vegetation "dead" due to lack of rainfall
Source: The authors, 2012.

Graph 10 below shows the relationship between the external air temperature and the average internal surface temperature of the measurement units. It can be seen that the temperature of the roofs varies in a very similar way to the outside air temperature, i.e. when there is an increase or decrease in the outside air temperature, the temperature of the inside surface of the roofs tends to follow the same behaviour.

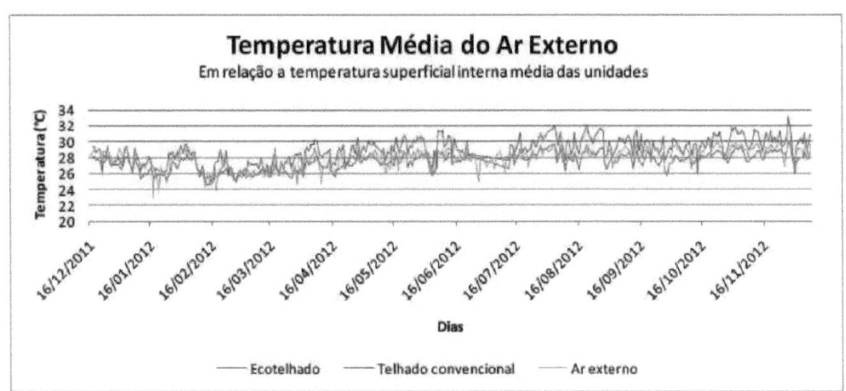

Graph 10: Relation of external air temperature to the average internal surface temperature of the measuring units
Source: The authors, 2014.

However, the UTC temperature is generally above the outside air temperature, on average 1

°C above, while the UET temperature is generally below the outside air temperature, on average 0.5 °C below. This highlights the characteristics of the different types of roof, since fibre cement has a greater capacity to absorb heat than green roofs.

Therefore, in general, the two measurement units, UTC and UET, suffer similar interferences from these climatic parameters, but the final results obtained show that they suffer these interferences with different intensities.

6.1 - Evaluation of the days with the highest temperatures

The period chosen for detailed analysis was 26 to 28 November. This period was chosen because it showed high average and peak temperatures, both for the internal air and the internal surface of the roof tiles, as well as expressively showing the specific behaviour of each type of roof tile.

With regard to the influence of climatic parameters, it was confirmed that solar insolation has a relationship with the temperature of the roof tiles and, consequently, the air inside the measuring units. On the days of the detailed study, which had high temperature values, insolation had an average incidence of 9h/day, while the average for the entire study period was 6.88h/day.

The relationship with humidity is also in line with what has already been discussed: humidity was low during this period, with an average of 72.7 per cent, while the average for the entire study period was 81.5 per cent. The average ambient air temperature was also high, with an average of 29.72°C, while the average for the whole period was 28°C. Furthermore, the 27th was the day with the highest average temperature for the whole period, with 30.27°C.

As for rainfall, there was no significant interference, and there was only rain on the 26th, which was not far off the average for the whole period. Therefore, this period was also chosen because it presented all this significant data. The table below shows the comparative data clearly.

Table 5 - Average daily weather data for the detailed study period (26, 27 and 28 November) and the entire period

Period	Parameters			
	Insolation (h/day)	Humidity (%)	Temperature (°C)	Precipitation (mm)
Detailed	9,07	72,67	29,72	9,27
Total	6,88	81,44	28,00	9,33

Source: The authors, 2014.

By analysing the data, it was possible to see the differences in thermal behaviour at UET and UTC, even when exposed to the same conditions. The figure below shows the air temperature values during the study period.

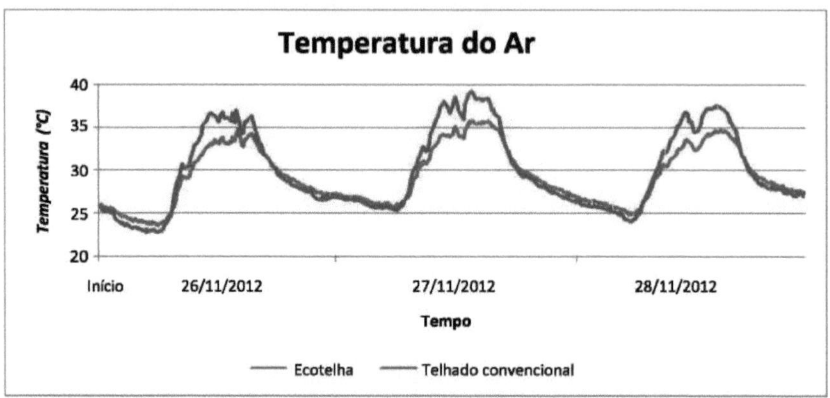

Graph 11 - Air temperature values between 26, 27 and 28 November 2012.

. Source: The authors, 2014.

In the ETU, the eco-roof had the characteristic of not accumulating heat and dissipating it more efficiently, acting as a thermal insulator and attenuating temperatures. This characteristic of the eco-roof led to lower temperature peaks at UET, ranging from 34.7°C to 35.8°C, while at UTC, maximum temperatures ranged from 37°C to 39.1°C on these days.

During the early hours of the morning, the temperature in the CTU was lower due to the thermal insulation caused by the ecotile in the ETU. Since the

fibre cement does not have this property, it loses heat more quickly, reaching lower levels at night.

In general, the air temperature was lower in the ETU, around 29.35°C, while in the CTU it was 29.91°C. The difference in the average internal air temperature between the units was 0.56°C, but at the most critical moments a difference of up to 3.9°C was reached.

It can also be seen that the temperature range is smaller in the ETU, i.e. the difference between the maximum and minimum temperature each day is smaller in the unit using the eco-roof. And this difference can be seen more clearly by analysing the surface temperature of the tiles (graph 12). The average temperature range over the three days was 25.47°C at UTC and 7.93°C at UET, when analysing the internal surface.

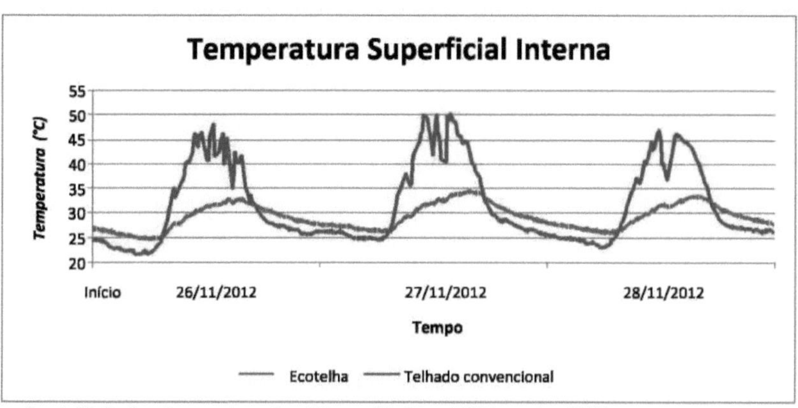

Graph 12: Surface temperature of roof tiles on 26, 27 and 28 November 2012.

. Source: The authors, 2014.

It can be seen that, when comparing the units, UTC reaches much higher values than UET, reaching up to 15.8°C more in this period. While UTC maximum temperatures reached 50.4°C, UET maximum temperatures were 34.6°C.

Therefore, the analysis led to the conclusion that the most striking characteristics identified in the eco-roof were a reduction in the average temperature inside the unit, a reduction in the temperature peaks in the unit and a reduction in the temperature range, as shown in the table below.

Table 6: Maximum, average and minimum air temperatures of the internal surfaces of the measurement units on 26, 27 and 28 November 2012.

	Air Temperature (°C)				Internal Surface Temperature (°C)			
Tile type	Average	Maximum	Minimum	Temperature range	Average	Maximum	Minimum	Temperature range
Ecotile (UET)	29,35	35,8	23,6	12,2	29,35	34,6	24,8	9,8
Conventional roof (UTC)	29,91	39,1	22,7	16,3	31,86	50,4	21,6	28,8
Difference	0,56	3,3	0,8	X	2,52	15,8	3,2	X

Source: The authors, 2014.

CHAPTER 7
CONCLUSION

This research allowed us to identify the thermal behaviour of a green roof and a conventional (fibre cement) roof in the city of Belém do Pará over a 12-month period - December 2011 to December 2012 - thus encompassing the rainy and less rainy periods in the Amazon.

As in the rest of Brazil, the use of eco-roofs on buildings in the Amazon region is a technique that is not widely used in urban centres compared to other countries where the technique is more widely used and encouraged. As such, the purpose of this study was to increase knowledge on the subject both in the Amazon region and in Brazil as a whole.

The results of the study showed that the use of eco-roofs considerably affects the temperature of a building's roof and, consequently, its interior.

Considerable characteristics could be noted with the use of the eco-roof compared to the conventional roof. The first is that there was an average temperature reduction of 1.5°C when comparing the internal surface temperature of the roof tiles, which generates a reduction of 0.26°C in the internal air temperature in the units. When analysing maximum air temperatures, this difference rises to 3° C, while the difference between maximum internal surface temperatures is an incredible 15.5 °C.

When analysing the influence of different meteorological parameters on the surface temperature of the measurement units, a significant relationship can be seen with the outside air temperature, humidity, solar insolation and rainfall.

The two measurement units, UTC and UET, suffer similar interference from these climatic parameters, but as the results show, it can be seen that they suffer these interferences with different intensities.

When evaluating the days with the most significant temperatures, it was possible to see the differences in thermal behaviour at UET and UTC, even when exposed to the same conditions.

Therefore, by using an eco-roof, it is easier to achieve an environment that provides thermal comfort, in addition to other benefits linked to the use of this type of roof, such as reducing surface run-off, reducing energy costs (cooling, heating...), attenuating heat islands, among others.

BIBLIOGRAPHICAL REFERENCES

ÁGUAS, P. N. M. **Thermal comfort,** Module of the Master's Degree course Instrumental Methods in Energy and the Environment, Instituto Superior Técnico. 2001.

AGUIAR, C. M. L. S.; FEDRIZZI, Beatriz. **Green roofs in social housing.** In: International Congress on Sustainability in Social Interest Housing, 2010, Porto Alegre.

ALMEIDA, M. A. M. de. **Natural cover and environmental quality: a contribution to a humid tropical climate.** Rio de Janeiro, Federal University of Rio de Janeiro , 2008. Available at : <http://teses2.ufrj.br/Teses/FAU_M/MarcoAntonioMilazzoDeAlmeida.pdf>. Accessed on: 12 May 2012.

Ambientebrasil: environmental portal. Available at: <http://www.ambientebrasil.com.br/>. Accessed on: Jan, 2012

ANDRADE, N. C. **Thermal performance of a green roof using Brachiaria humidicola grass in the city of São Carlos, SP.** 2007. 113 f. Master's dissertation. Department of Civil Engineering, Federal University of São Carlos, 2007.

ARAÚJO, G. M. **The degradation of water resources in the mata fome river basin, Belém-PA: a consequence of the urbanisation process. Belém - PA, 2007.** Final paper for a bachelor's degree in Social Work - Federal University of Pará. 2007.

ARAÚJO, M. A. **Modern sustainable construction.** Available at: <http://www.idhea.com.br/pdf/moderna.pdf>. Accessed on: 12 July 2011.

ARAÚJO, S. R. **As funções dos Telhados Verdes no Meio Urbano, na Gestão e no Planejamento de Recursos Hídricos.** Monograph Presented to the Forestry Institute of the Federal Rural University of Rio de Janeiro, Seropédica, RJ, 2007;

Brazilian Metal Construction Association. Technical Manual: Steel roof tiles. Sep, 2009. 36 p.

AZEREDO, H.A. de. **The building up to its roof.** Edgard Blucher Publishing House. São Paulo, 1997.

AZEVEDO, H. A. (1977). **O Edifício até sua Cobertura. 2 ed.** São Paulo Brazil: Editora Edgar Blucher Ltda., 1977.

BACOVIS, T. M. **Hydrological performance of an extensive green roof prototype.** 2010. 52 f. Final paper for the degree course in Environmental Engineering - Pontifical Catholic University of Paraná, Curitiba.

BEYER, P. **Measuring the Thermal Performance of Eco-Tiles.** Date not given. Vapour and

Refrigeration Laboratory, UFRGS. Porto Alegre, RS. Available at: <http://www.ecotelhado.com.br/InformacoesInterna/Desempenho%20termico.pdf>. Accessed on 05 June 2012.

BROWN, G. Z.; DEKAY, M. **Sun, wind & light: Strategies for architectural design.** 2ª .ed. Porto Alegre: Bookman, 2004, 415p.

CARDOSO, F. F. **Roof coverings: lecture notes.** Nov, 2000. São Paulo. Available at: <http://pcc2436.pcc.usp.br/Textost%C3%A9cnicos/coberturas/ApostilaCoberturaPCC436Ano2000.pdf>. Accessed on 05 June 2012.

CASTLETON, H. F., V.STOVIN, BECK, S. B. M., & DAVISON, J. B. (2010). **Green roofs: building energy savings and potential for retrofit.** Energy and Building 42, 1582-1591.

CATUZZO, H. ; Lombardo, Magda Adelaide . **Green roofs: possibilities for reducing the heat island in the metropolis of São Paulo.** In: XII SIMPÓSIO NACIONAL DE GEOGRAFIA, 2011, Belo Horizonte. Science and Utopia: towards a geography of the possible. Belo Horizonte : SIMPURB, 2011. v. 12. p. 1-12.

CAVALCANTI, I. F. A.; FERREIRA, N. J.; SILVA, M. G.A.J.; SILVA DIAS, M. A. F. **Weather and Climate in Brazil**. 1 ed. Sao Paulo: Oficina de Textos, 2009, v. 1.

CORRÊA, C. B. **Green roofs: the ecological roof.** Revista CREA/RS, issue: 34, p 28, June 2007. Available at: <http://www.crea-rs.org.br/site/archivo/revistas/ed34.pdf>. Accessed on: 12 July 2012.

COSTA, A.C.L., OLIVEIRA, M.C.F. **Influência das atividades urbanas no comportamento da temperatura do ar na cidade de Belém-Pa.** Proceedings of the IX Brazilian Congress of Meteorology, vol.2, 1996.

COSTA, A. C. L. da. MATTOS, A. **Correlation analysis between air temperature and the main urban elements in the city of Belém - PA.** Department of Meteorology - Federal University of Pará.

CREA/RJ. **The hanging gardens of Babylon.** CREA/RJ magazine, issue: 75, pages 58 to 60, December. 2008. Available at: <http://app.crea-rj.org.br/portalcreav2midia/documentos/revista75.pdf>. Accessed on: 12 July 2012.

DA SILVA, L. S. da. **Evolution of roofing systems used in Brazil.**

2005. 124 f. Final paper for the degree course in Civil Engineering with an emphasis on the environment - Anhembi Morumbi University, São Paulo - SP.

DE SOUZA, E.B. et al. **Seasonal precipitation over the eastern Amazon during the rainy season: Observations and regional simulations with RegCM3.** Revista Brasileira de Meteorologia, v.24, n.2, 111-124, 2009

Eco telhado: Available at: <http://www.ecotelhado.com.br>. Accessed on: 12 Nov. 2011.

FERREIRA, M. de F. **Teto verde: o uso** de **coberturas vegetais em Edificações.** 2007. Rio de Janeiro - RJ. 2007. Available at: <http://www.puc-rio.br/pibic/relatorio_resumo2007/relatorios/art/art_manoela_de_freitas_ferreira.pdf> Accessed on: 29 Mar. 2012.

FROTA, A. B.; SCHIFFER, S. R. **Manual de Conforto térmico: arquitetura, urbanismo.** 5ª edition. São Paulo: Studio Nobel, 2001.243 p.

GOMES, H. M. **Society/nature relations and the economic valuation of nature: the case of the state forest - Cabedelo - PB.** 1999. 123 f. Dissertation (Master's Degree in Environment and Development) - Federal University of Paraíba, João Pessoa. 1999.

GOULART, S.V.G., BARBOSA, M.J., PIETROBON, C.E., BOGO, A., PITTA, T. **Bioclimatology applied to building design for thermal comfort.**

Florianópolis: Construction Research Centre/UFSC, 1994 (internal report no. 02/94).

GOUVEIA, L. V. de. TETO VERDE: **An ecological proposal to improve environmental comfort through the use of** green **roofs in buildings.** 2007. Rio de Janeiro - RJ. 2007. Available at: <http://www.puc- rio.br/pibic/relatorio_resumo2008/relatorios/ctch/art/art_lauravg.pdf> Accessed on: 29 Mar. 2012.

Greenroofs.com: The Resource Portal for Green Roofs. Available at: <http://www.greenroofs.com/>. Accessed on: 12 Nov. 2011.

GRÚN, M. **Ethics and environmental education: the necessary connection.** 11ª Edition. Campinas: Papirus, 2007. 120 p.

GURGEL, R.A.G. **Main species and varieties of grass.** In: SYMPOSIUM ON GRASS, 1, 2003, Botucatu. Production, implementation and maintenance: Proceedings. Botucatu: Department of Natural Resources, Faculty of Agronomic Sciences, São Paulo State University, 2003.

HENEINE, M. C. Al. de S. **Green cover.** Belo Horizonte, Federal University of Minas Gerais, 2008. Available at: < http://www.cecc.eng.ufmg.br/trabalhos/pg1/Monografia%20Maria%20Cristina%20Almeida.pdf>. Accessed on 26 February 2012.

HITOSH, C. . 2000. **The Sky the Limit look Japan: Greening the Cities - Japan's pioneers in energy Conservation,** 46(545), pp. 6-15.

IBGE. 2010 Population Census. 2010 Population Census. Brazilian Institute of Geography and Statistics (IBGE). Available at: <http://www.ibge.gov.br/home/>. Accessed on 28 Feb. 2012.

IMBIRIBA JÚNIOR, M.; COSTA, F. R.. **Water Resources: The case of the Bolonha and Água Preta lakes in the Metropolitan Region of Belém, Pará.** In: 33ª ASSEMAE National Assembly, 2003, Santo André - SP, 2003.

National Meteorological Institute (INMET) Available at: < http://www.inmet.gov.br/portal/>. Accessed on: 29 Oct, 2014

INMET. 1992. Climatological Normals. 155p.

KOERICH, D, W. **The evolution of the use of the garden slab.** Cascavel, 2010. Final paper for the degree course in Architecture and Urbanism - Assis Gurgacz College, Cascavel.

LAMBERS, R; XAVIER, A. A. P. **Conforto Térmico e Stress Térmico.** Florianópolis: Ed: Universitária/UFSC, 2002.

LEE, A. **The cost of alternatives for replacing asbestos cement roofing.** São Paulo, Polytechnic School of the University of São Paulo. 12p. 2000. Available at: <http://www.web-resol.org/textos/Telhados de cimento amianto.pdf>. Accessed on 26 February 2012.

Complementary Law No. 027, published in the Municipal Official Gazette on 19 October 1995

LENOTTI, J. **Work on Building Organisation and Installations I.** August 2010. Available at: < http://www.ebah.com.br/content/ABAAABG6gAG/trabalho- coverage>. Accessed 26 Feb. 2012.

LIMA, I. G; BARROCA, B. B.; D'OLIVEIRA, P. S. **Influence of the ecological roof with green plants on environmental comfort.** In: ENCONTRO INTERNACIONAL DE PRODUÇÃO CIENTÍFICA CESUMAR, 6., 2009. Maringá. *Proceedings...* Maringá, 2009.

LOGSDON, N. B. **Wooden structures for roofs, from the perspective of NBR 7190/1997.** Faculty of Forestry Engineering, Federal University of Mato Grosso. Cuiabá, MT. 2002.

LOPES, D. A. R. 2007. **Analysis of the thermal behaviour of a light green roof and different roofing systems.** Master's dissertation. Postgraduate Programme in Environmental Engineering Sciences, São Carlos School of Engineering, University of São Paulo. São Carlos - SP.

LOPES, E. L. N.; Fernandes, A. R.; Grimaldi, C.; Ruivo, M. L. P.; Rodrigues, T. E.; Sarrazin, M.;

Boletim do Museu Paraesense Emílio Goeldi, Ciências Naturais 2006, 1, 127.

MARY W. et. al. **Green roofs: a potential tool for generating income in areas of social fragility.** 9° Encontro Nacional de Ensino de Paisagismo em Escolas de Arquitetura e Urbanismo no Brasil - ENEPEA- Curitiba, Paraná - 2008.

MELLO, G. B. O.; COSTA, M. D. P.; ALBERTI, M. S.; FREITAS FILHO, R. D. G. **Study of the implementation of a green roof at the mechanical engineering faculty.** Revista Ciências do Ambiente On-Line, São Paulo, v. 6, n. 2, Dec. 2010. Available at: <http://sistemas.ib.unicamp.br/be310/viewissue.php?id=13>. Accessed on: 12 July 2011.

MENDONÇA, F.; DANNI-OLIVEIRA, I. M. **Climatologia: notções básicas e climas do Brasil.** São Paulo: Oficina de Texto, 2007. 206 p.

MORAIS, C.S.; RORIZ, M. **Comparison between the thermal performance of garden roofs and common slabs in guardhouses.** In: 7° National Meeting on Comfort in the Built Environment - ENCAC. PUC Curitiba/2003.

MORAIS, C.S.; RORIZ, M. **Temperatures in a prototype building with a garden roof: São carlos, SP.** In: VI National Meeting on Comfort in the Built Environment (ENCAC) and IV Latin American Meeting on Comfort in the Built Environment (ELAC), 2005.

NASCIMENTO, L., 2010. **The wave of green roofs.** O eco, June 2010. Available at: <http://www.oeco.com.br/reportagens/24075-a-onda-dos-telhados- ecologicos>. Accessed on: 12 Jan. 2012.

NASCIMENTO W. C. **Coburas verdes no Contexto da Região Metropolitana de Curitiba: barreiras e potenciaisidades.** Curitiba, 2008. Dissertation (Master's in Civil Construction) - Postgraduate Programme in Civil Construction, Federal University of Paraná.

NIMER, E. **Climatologia do Brasil.** 2.ed. Rio de Janeiro: IBGE Foundation, 1989.

NOGUEIRA, M. C. J. A.; DURANTE, L. C.; NOGUEIRA, J. S. **Conforto térmico na escola pública em Cuiabá-MT: estudo de caso.** *Revista Eletrónica do Mestrado em Educação Ambiental.* V. 14, jan/jun. 2005. Available at: <http://www.remea.furg.br/>. Accessed on: 12 Dec. 2011.

OLIVEIRA, E. W. N. de. **Green roofs for social housing: rainwater retention and thermal comfort.** 2009. 86f. Dissertation (Master's in Environmental Engineering) - Faculty of Engineering, Rio de Janeiro State University, Rio de Janeiro, 2009.

OLIVEIRA, M. do C. F. de; ABREU, J. R. R.; BARROS, A. N.F.; BEZERRA, A. C. N.; MACHADO,

J. R. S.; NETO B. S. **Monthly and annual rainfall trends in Belém-PA, from 1896 to 2000 (105 years).** In: Brazilian Congress of Meteorology, 12, 2002, Foz do Iguaçu - Paraná.

OLIVEIRA, L. L. FONTINHAS, R. L. LIMA, A. M. M. LIMA, R. J. S. **Maps of the climatological parameters of the state of Pará: humidity, temperature and insolation, annual averages.** In XIII Brazilian Congress of Meteorology, 2002.

ORTIZ, R. **Reflexões sobre a pós-modernidade: o exemplo da arquitetura.** *Revista Brasileira de Ciências Sociais,* n. 20, October 1992.

PALHA, P. **Garden roofs: a solution for cities?** APH Magazine, issue: 106, p 26 to 31, July/August/September. Available at: < http://www.neoturf.pt/ficheiros/file/ArtigoPauloPalhanaAPH.pdf>. Accessed on: 15 March 2012.

PARÁ: Government Portal. Available at: <http://www.pa.gov.br/> Accessed on: 29 Mar. 2012.

PECK, S., et al. **Greenbacks from Green Roofs: Forging a New Industry in Canada.** Canada Mortgage and Housing Corporation. Mar, 1999.

PINHEIRO, J. A. N. **Urban heat island.** 2008. Belém - PA. Federal University of Pará. Available at < http://www.webartigos.com/artigos/ilha-de-calor- urbana/10654/> Accessed on: 05 Apr. 2012.

ROCHA, E. J. P. **Moisture Balance and Influences of Surface Boundary Conditions on Amazonian Precipitation. 2001.** 210 f. (Doctoral Thesis) - National Institute for Space Research, INPE, São Paulo, 2001.

ROLIM, P.A.M.; SANTOS, D.M.; ROCHA, E.J.P. **Precipitation Variability in the Amazon: Socioeconomic Implications.** In: XIV Brazilian Congress of Meteorology, 2008, Florianópolis. Electronic Proceedings. Florianópolis: SBMET, 2006. Available at: <http://www.cbmet.com/cbm-files/14- 1b5c3e3f6e34f1ca974cb06dd7bf1c6c.pdf>. Accessed on 18 May 2012.

RUAS, A. C. **Avaliação de conforto térmico - contribuição à aplicação prática das normas internacionais.** Campinas, 1999. Master's dissertation - School of Civil Engineering, State University of Campinas.

SEROA da MOTTA, R., **Manual para Valoração Económica de Recursos Ambientais,** IPEA/MMA/PNUD/CNPq, Rio de Janeiro, 1997.

SEVEGNANI, K.B.; GHELFI FILHO, H.; SILVA, LJ.O. **Comparison of various roofing materials using thermal comfort indices.** Scientia Agrícola, Piracicaba, v.51, n.1, p.1-7, 1994.

SIQUEIRA, C. **Environmental comfort, a challenge for architects.** Available at Forum da Construção< http://www.forumdaconstrucao.com.br/conteudo.php?a=4&Cod=800> Accessed on: 30 Jan. 2012.

STRAPASSON, D. C.; Freitas, M. C. D.; DOS SANTOS, A. **Comparative Study of the Energy Consumption of Conventional and Light Green Roofs in a Building, Ponta Grossa,** In: 5º Encontro de Engenharia e Tecnologia dos Campos Gerais (EETCG), 2010.

TINOCO, J. E. L.; **Traditional roofs: pathologies, repairs and maintenance.** In: III SYMPOSIUM ON ADVANCED TECHNIQUES IN THE CONSERVATION OF CULTURAL PROPERTY. 2006, Olinda - RE.

TOMAZ, 2005. **Green roof.** 2005. Chapter 10. Available at < http://www.scribd.com/doc/73596806/Capitulo10-Telhadoverde> Accessed on: 12 July 2012.

TOMAZ, 2008. **Green roof.** 2008. In: Rainwater handling course. Chapter 51. Available at < http://www.pliniotomaz.com.br/downloads/livro_poluicao_difusa/capitulo51.pdf> Accessed on: 12 July 2012.

VACILIKIO, D. V.; FLEISCHFRESSER. L. **Comparison between Green and Conventional Roofs in Internal Temperatures of Environments.** In: Scientific Congress of the Centre-Western Region of Paraná, 4.M 2011, Campo Mourão. Proceedings.

VECCHIA, F. **Cobertura Verde Leve (CVL): Ensaio Experimental,** Maceió, In: VI Encontro Nacional de Conforto no Ambiente Construído (ENCAC) e IV Encontro Latino-americano sobre Conforto no Ambiente Construído (ELACAC), 2005.

WONG, N. H. et al. 2003. **The effects of rooftop garden on energy consumption of a commercial building In Singapore.** Energy and Buildings, v.35, n°. 4, p. 353- 364, 2003.

WONG, M. **Environmental benefits of green roofs.** Available at <http://www.nea.gov.sg/cms/sei/PSS23slides.pdf> Accessed on: 12 Jul. 2012.

XAVIER, A. A. P. **Thermal comfort conditions for high school students in the Florianópolis region.** 1999. 209 f. Dissertation (Master's in Civil Engineering) - Federal University of Santa Catarina, Florianópolis, 1999.

YAZIGI, W. **The technique of building.** São Paulo: SindusCon-SP, Editora PINI, 1998.

ZAMPIER, J. F.jMIRANDA, G. M. **Survey of methodologies proposed for the economic valuation of environmental goods.** *Revista Eletrónica Lato Sensu. v.* 2, n. 1, jul. 2007. Available at: <http://web03.unicentro.br/especializacao/Revista_Pos/Lato_Sensu.htm> Accessed on: 19 July 2012.

Appendix A- Average daily data obtained from the entire period studied (starting on 15/12/2011 and ending on 09/12/2012)

	Ecotile						Conventional					
Day	Max Air	Min Air	Medical Air	Ts Max	Ts Min	Ts Méd	Max Air	Min Air	Medical Air	Ts Max	Ts Min	Ts Méd
15/12/2011	31,80	26,10	29,28	31,20	27,10	29,55	34,10	25,40	29,34	43,80	24,30	30,01
16/12/2011	32,80	23,90	27,84	31,80	25,10	28,13	35,80	23,30	28,09	47,90	22,20	29,47
17/12/2011	33,00	23,10	27,64	32,30	24,60	28,18	35,10	22,10	27,78	46,00	20,80	29,29
18/12/2011	32,10	23,60	27,44	31,30	25,00	27,74	34,90	22,80	27,57	46,60	21,40	28,70
19/12/2011	32,40	24,30	27,86	31,20	25,00	27,83	34,80	23,70	28,09	43,60	22,60	29,02
20/12/2011	32,40	24,00	27,31	31,50	25,10	27,47	35,40	23,60	27,59	49,20	22,70	29,00
21/12/2011	31,00	23,80	26,04	28,60	24,20	25,97	33,60	23,00	25,83	43,70	21,60	26,41
22/12/2011	31,70	23,90	27,46	30,50	24,70	27,45	34,20	22,10	27,38	42,30	23,20	28,38
23/12/2011	31,00	24,00	27,43	30,10	25,30	27,75	32,90	23,20	27,46	40,90	22,00	28,04
24/12/2011	32,10	23,70	27,67	31,40	24,90	28,06	34,70	23,00	27,96	45,40	21,90	29,42
25/12/2011	31,90	23,50	26,64	30,90	24,90	27,21	34,40	22,80	26,70	43,10	22,00	27,53
26/12/2011	31,50	23,50	26,78	30,30	24,50	27,14	33,60	22,80	26,73	42,00	21,30	26,89
27/12/2011	30,40	24,50	26,10	29,20	24,60	26,19	31,90	24,10	25,91	37,30	23,20	25,93
28/12/2011												
29/12/2011												
30/12/2011	32,10	23,70	27,03	30,80	25,00	27,07	34,70	22,90	27,05	45,90	21,70	28,00
31/12/2011	30,90	23,70	26,63	29,50	24,60	26,44	33,00	23,00	26,64	42,30	21,60	27,03
01/01/2012	31,20	24,20	27,14	29,60	25,00	27,16	33,60	23,40	27,18	43,60	22,20	27,71
02/01/2012	31,60	24,60	27,74	30,70	25,40	27,92	34,00	24,00	28,01	43,10	23,00	29,24
03/01/2012	32,30	24,20	27,85	31,50	25,20	28,16	34,40	23,60	28,06	41,90	22,70	29,28
04/01/2012	31,20	24,70	27,23	30,20	25,70	27,51	33,00	24,20	27,18	42,20	23,40	27,57
05/01/2012	30,60	23,70	26,40	30,10	24,40	26,79	32,10	23,00	26,34	38,40	22,00	26,76
06/01/2012	30,80	23,40	26,38	28,80	23,80	26,22	33,40	22,70	26,39	41,90	21,60	26,61
07/01/2012	33,30	23,70	27,91	31,60	24,70	27,96	35,50	22,90	28,06	44,50	21,50	29,10
08/01/2012	33,40	24,00	26,95	31,90	25,40	27,15	35,50	23,20	26,93	46,80	22,00	27,86
09/01/2012	30,60	24,30	26,96	28,50	24,60	26,31	32,80	23,80	27,18	41,30	22,80	27,51
10/01/2012	31,70	23,50	25,81	29,10	23,80	25,38	34,90	22,50	25,86	45,70	21,40	26,32
11/01/2012	31,60	23,60	26,13	29,00	23,80	25,72	34,80	23,00	26,37	44,90	22,00	27,04
12/01/2012	31,10	23,60	26,50	28,60	24,10	26,04	33,80	22,90	26,67	43,60	21,70	27,04
13/01/2012	31,50	24,10	26,71	28,70	24,50	25,87	33,90	23,60	26,95	42,10	22,50	27,60
14/01/2012	31,70	23,30	27,17	29,30	23,90	26,37	34,40	22,60	27,44	42,80	21,30	28,14
15/01/2012	31,40	24,80	27,04	30,10	25,50	26,85	33,40	24,00	26,95	42,00	23,20	27,31
16/01/2012												
17/01/2012	29,40	22,50	25,85	27,80	22,80	25,31	31,10	22,00	26,11	36,10	21,30	26,65
18/01/2012	30,80	23,20	26,11	28,80	23,90	25,93	33,40	22,30	25,71	44,00	21,30	26,45
19/01/2012												
20/01/2012												
21/01/2012												
22/01/2012	30,30	23,70	25,91	28,30	24,00	25,77	32,70	23,20	25,87	42,40	22,10	26,15
23/01/2012	30,40	23,50	25,39	28,20	23,70	25,26	33,00	23,00	25,54	43,70	22,20	26,29
24/01/2012	31,50	23,80	26,48	29,00	24,00	25,88	34,30	23,40	27,01	44,30	22,60	28,51
25/01/2012	32,40	23,40	26,88	29,60	24,00	26,22	35,60	22,70	27,23	44,30	21,40	28,36
26/01/2012	33,30	24,50	27,52	31,20	25,20	27,51	36,10	23,80	27,63	46,20	22,80	28,71
27/01/2012	31,60	24,20	27,10	30,60	25,60	27,31	33,20	23,40	27,04	44,10	22,10	27,86
28/01/2012	31,90	23,60	27,12	30,00	24,20	26,62	34,60	22,80	27,43	44,70	21,60	28,66
29/01/2012	31,30	24,10	27,47	29,40	24,80	26,98	33,50	23,40	27,65	42,50	22,20	28,50
30/01/2012	32,50	24,00	27,83	31,40	25,10	28,01	34,30	23,20	28,03	44,30	22,10	29,43
31/01/2012	32,70	23,60	27,39	31,70	24,90	27,62	35,20	22,60	27,45	45,80	21,30	28,77
01/02/2012	33,70	23,90	27,96	31,90	24,70	27,86	36,40	23,10	28,23	46,80	21,80	29,60
02/02/2012	32,90	24,50	28,05	31,90	25,80	28,39	34,90	23,90	28,21	45,10	23,00	29,55
03/02/2012	34,00	23,50	27,35	32,40	24,70	27,81	36,40	22,80	27,51	45,80	21,70	28,68
04/02/2012	32,70	24,00	27,21	31,70	25,00	27,70	34,90	23,50	27,26	45,20	22,50	28,34

05/02/2012	33,10	23,90	27,49	32,30	25,10	27,82	35,60	23,20	27,63	44,70	22,00	28,77
06/02/2012	33,30	24,10	27,64	31,70	25,20	28,08	35,60	23,40	27,68	44,50	22,00	28,74
07/02/2012	32,10	24,20	26,51	30,70	25,00	26,78	34,60	23,40	26,43	46,70	22,40	26,87
08/02/2012	29,70	24,20	26,00	27,90	24,40	25,61	31,70	23,70	26,09	38,50	22,90	26,29
09/02/2012	29,00	24,30	25,97	27,50	24,60	25,75	30,50	23,90	26,01	33,90	23,20	26,12
10/02/2012	29,50	24,80	26,62	28,00	25,30	26,28	31,40	24,30	26,67	38,50	23,70	26,93
11/02/2012	30,50	23,40	25,20	29,10	23,60	25,29	32,70	22,80	25,14	39,10	22,00	25,49
12/02/2012	30,00	22,40	24,88	27,80	22,90	24,63	32,20	21,70	25,00	39,20	20,70	25,47
13/02/2012	29,50	22,40	24,73	27,70	22,70	24,51	32,00	21,80	24,93	42,30	21,10	25,51
14/02/2012	30,10	23,60	25,17	28,40	23,70	25,11	32,90	23,20	25,24	41,90	22,40	25,76
15/02/2012	28,80	22,60	24,74	27,20	23,20	24,89	30,90	22,10	24,73	34,90	21,60	24,87
16/02/2012	31,10	23,00	25,87	29,20	23,80	25,82	33,60	22,30	26,22	41,60	21,10	27,35
17/02/2012												
18/02/2012	30,40	23,00	25,94	28,10	23,20	25,58	33,20	22,50	26,44	41,40	22,00	27,61
19/02/2012	32,00	24,00	26,60	29,40	24,60	26,34	35,10	23,50	27,16	48,70	22,80	29,05
20/02/2012	30,20	23,70	26,14	28,50	24,40	26,02	33,10	23,00	26,27	43,70	21,90	26,78
21/02/2012	31,00	23,30	25,98	29,20	24,00	26,16	34,10	22,60	26,07	45,40	21,40	26,91
22/02/2012	33,10	22,70	26,77	31,00	23,60	26,59	36,20	21,80	27,20	47,30	20,60	28,94
23/02/2012	30,90	23,60	25,78	28,80	24,20	25,85	33,20	22,80	25,66	41,10	21,60	25,75
24/02/2012	30,80	23,70	26,34	28,70	24,40	26,03	33,30	23,00	26,34	42,60	21,90	26,74
25/02/2012	28,50	23,80	26,07	27,10	24,30	25,80	29,80	23,40	26,24	33,70	22,90	26,64
26/02/2012	28,20	24,10	25,44	26,90	24,60	25,54	29,10	23,60	25,22	31,80	22,90	25,21
27/02/2012	29,00	23,60	25,20	27,20	24,20	25,37	31,50	22,80	25,05	44,00	21,70	25,23
28/02/2012	29,50	23,10	25,59	27,90	23,80	25,73	31,60	22,10	25,52	38,80	21,10	25,78
29/02/2012	30,60	23,10	25,84	28,30	23,90	25,83	33,00	22,40	25,84	41,50	21,30	26,47
01/03/2012	30,70	23,50	25,75	28,60	24,20	25,65	33,40	22,80	25,68	42,10	21,80	26,12
02/03/2012	30,60	23,60	25,90	28,60	24,20	25,86	32,90	22,90	25,91	43,80	21,80	26,53
03/03/2012	30,30	23,10	25,67	28,20	23,80	25,52	32,60	22,30	25,61	41,90	21,20	26,06
04/03/2012	29,90	23,50	26,33	28,10	24,00	25,82	32,40	23,00	26,65	40,30	22,20	27,52
05/03/2012	31,60	24,30	26,08	29,20	24,60	25,84	34,70	23,80	26,21	48,40	23,00	27,03
06/03/2012	31,60	23,70	25,80	28,90	24,40	25,64	34,90	22,90	25,94	47,00	21,80	27,05
07/03/2012	30,80	23,50	25,92	29,00	23,90	25,72	33,90	22,70	26,08	43,00	21,60	27,05
08/03/2012	31,00	23,40	25,94	29,40	24,00	25,96	34,10	22,80	26,07	46,40	22,10	27,27
09/03/2012	30,70	23,30	26,03	28,90	24,00	26,02	33,50	22,50	26,01	43,40	21,40	26,62
10/03/2012	31,60	23,80	26,47	30,00	24,40	26,45	34,90	23,20	26,63	46,30	22,50	28,03
11/03/2012	31,10	24,00	26,47	29,40	24,90	26,53	34,00	23,30	26,49	46,30	22,40	27,20
12/03/2012	29,40	24,40	26,07	28,10	25,10	26,21	31,60	23,70	25,98	39,10	23,00	26,25
13/03/2012	30,40	23,80	26,32	29,10	24,60	26,38	32,60	23,20	26,34	43,50	22,20	27,14
14/03/2012	31,00	23,70	25,89	28,30	24,40	25,71	33,60	23,10	25,89	42,40	21,90	26,29
15/03/2012	29,80	22,80	25,74	27,90	23,50	25,59	31,60	22,10	25,68	38,40	20,80	26,08
16/03/2012	31,90	23,60	26,73	28,90	24,30	26,42	34,90	22,70	26,84	45,70	21,60	27,81
17/03/2012	33,30	24,00	27,75	31,00	24,90	27,16	36,20	23,20	27,95	46,40	22,20	29,17
18/03/2012	31,90	23,90	26,48	30,50	24,80	26,64	34,10	23,10	26,39	42,90	22,00	26,96
19/03/2012	30,30	23,90	25,87	28,30	24,50	25,79	32,50	23,30	25,83	39,50	22,20	26,13
20/03/2012	31,70	23,40	26,47	29,10	24,10	26,27	35,40	22,70	26,65	47,50	21,80	27,51
21/03/2012	31,80	23,90	25,96	29,20	24,70	25,95	34,70	23,00	25,90	47,40	22,00	26,54
22/03/2012	31,90	24,00	26,56	29,10	24,50	26,30	34,70	23,40	26,62	44,20	22,50	27,41
23/03/2012	31,40	23,90	26,28	29,00	24,60	26,08	34,40	23,10	26,45	46,10	21,90	27,59
24/03/2012	31,50	23,30	26,23	29,60	22,60	25,62	34,00	23,20	26,66	43,50	22,30	28,07
25/03/2012	31,40	23,50	25,79	29,10	23,90	25,48	35,00	22,90	25,93	47,70	22,00	26,81
26/03/2012	31,90	23,90	26,23	29,60	24,20	25,93	34,90	23,40	26,29	46,80	22,50	27,07
27/03/2012	31,70	24,30	26,89	29,00	24,80	26,55	34,70	23,70	27,06	49,20	22,80	28,38
28/03/2012												
29/03/2012	31,40	23,40	26,85	28,70	23,80	26,11	34,10	22,90	27,12	45,60	22,10	28,44
30/03/2012	31,90	24,40	27,17	29,80	25,10	26,83	35,20	23,80	27,33	48,00	23,00	28,54
31/03/2012	30,30	24,40	25,77	28,80	23,70	25,55	31,80	23,90	25,90	40,00	23,10	26,44
01/04/2012	32,30	24,50	27,53	30,30	24,90	27,21	35,00	23,80	27,68	48,50	22,80	28,88
02/04/2012	33,10	24,40	27,74	31,70	25,70	27,85	36,30	23,50	27,82	50,50	22,50	29,38
03/04/2012	33,40	24,50	27,34	31,20	25,10	27,05	37,10	23,70	27,48	51,00	22,80	28,85
04/04/2012	33,20	23,80	27,42	31,90	24,80	27,39	36,50	22,90	27,65	49,70	21,70	29,62
05/04/2012	33,50	24,00	27,42	32,40	25,00	27,38	36,20	23,20	27,69	50,10	22,10	29,59
06/04/2012	33,60	24,20	28,19	32,10	24,90	27,74	36,70	23,50	28,54	46,90	22,40	30,24
07/04/2012	33,00	25,20	28,29	31,50	26,40	28,34	36,20	24,50	28,54	48,60	23,30	30,12
08/04/2012	32,80	24,10	27,08	31,60	25,00	27,38	35,60	23,50	27,09	51,60	22,60	28,41
09/04/2012												
10/04/2012	31,60	23,70	26,77	29,90	24,30	26,59	34,80	23,10	26,91	49,30	21,90	28,16

11/04/2012	32,20	23,90	26,99	29,80	24,70	26,65	35,70	23,20	27,15	49,50	22,00	28,50
12/04/2012	32,70	23,70	27,01	30,90	24,50	26,91	35,60	23,00	27,14	49,70	21,90	28,53
13/04/2012	32,10	24,00	27,57	30,30	24,90	27,66	34,80	23,20	27,73	49,40	22,00	29,07
14/04/2012	31,80	24,30	26,05	29,60	24,90	26,33	35,40	23,80	26,04	50,10	23,00	26,64
15/04/2012	31,50	23,40	25,96	29,70	23,90	25,86	34,90	22,70	26,15	49,30	21,70	27,37
16/04/2012	30,70	23,30	25,68	28,40	23,80	25,60	34,50	22,80	25,96	46,80	22,10	27,07
17/04/2012	31,90	23,60	26,36	29,40	24,30	26,29	35,70	22,80	26,54	49,50	21,80	27,85
18/04/2012	32,40	23,60	26,91	29,60	24,30	26,68	36,60	22,90	27,25	52,80	21,90	29,16
19/04/2012	32,20	23,60	26,99	29,70	24,50	26,78	35,20	22,90	27,18	49,00	21,70	28,65
20/04/2012	31,80	24,20	27,35	29,70	24,90	27,17	34,90	23,60	27,73	50,00	22,60	29,57
21/04/2012	30,60	25,00	26,12	29,80	25,90	26,71	32,40	24,40	26,02	42,60	23,80	26,92
22/04/2012	32,40	23,70	26,78	31,10	25,10	27,34	34,80	22,90	26,55	44,60	21,80	27,24
23/04/2012	32,50	24,20	26,73	31,00	25,10	26,89	35,70	23,60	26,78	50,70	22,60	28,01
24/04/2012	31,70	23,90	27,21	29,60	24,50	26,88	35,10	23,30	27,60	50,00	22,20	29,49
25/04/2012	32,40	24,20	27,42	30,40	25,10	27,31	36,00	23,60	27,70	50,30	22,60	29,40
26/04/2012	32,10	24,50	27,15	30,10	25,40	27,21	35,20	23,80	27,22	50,00	22,60	28,50
27/04/2012	32,60	24,40	28,02	31,10	25,30	28,22	35,70	23,70	28,36	50,50	22,50	30,45
28/04/2012	32,20	24,60	27,10	31,30	25,70	27,57	34,60	23,80	27,06	46,20	22,70	28,00
29/04/2012	33,10	23,80	27,18	31,50	24,80	27,30	36,00	23,10	27,34	47,50	22,00	28,79
30/04/2012	33,00	23,90	27,48	31,60	25,20	28,05	36,40	22,80	27,68	52,20	21,70	29,41
01/05/2012	33,40	23,90	27,02	32,10	25,50	27,64	36,60	23,10	27,14	52,80	21,90	28,66
02/05/2012	33,90	23,50	27,55	32,30	24,70	27,95	37,80	22,80	27,92	51,60	21,60	30,04
03/05/2012	33,30	24,50	27,93	32,30	25,80	28,53	36,70	23,90	28,13	50,60	22,90	29,80
04/05/2012	33,50	24,40	27,73	32,60	25,70	28,52	37,00	23,60	27,89	50,70	22,60	29,67
05/05/2012	34,40	23,70	27,55	33,40	25,30	28,20	38,40	22,90	27,77	53,90	21,80	29,84
06/05/2012	33,30	23,60	27,22	32,00	24,80	27,78	36,30	22,90	27,56	49,70	21,80	29,47
07/05/2012	33,80	23,50	27,99	32,50	24,90	28,23	37,50	22,70	28,62	53,60	21,40	32,11
08/05/2012												
09/05/2012	32,40	22,90	26,90	31,30	23,90	27,05	35,80	22,10	27,14	49,00	21,10	28,78
10/05/2012	33,20	23,50	27,22	31,90	24,70	27,52	37,10	22,70	27,42	51,40	21,50	29,18
11/05/2012	32,30	23,80	26,40	31,20	25,10	27,03	35,90	22,90	26,42	51,30	21,70	27,68
12/05/2012	32,50	23,60	27,20	30,90	24,60	27,25	36,20	22,80	27,42	48,20	21,80	28,77
13/05/2012	33,20	24,10	26,85	31,70	25,40	27,60	36,80	23,20	26,92	53,20	22,30	28,40
14/05/2012	33,70	23,70	27,34	32,30	25,10	27,48	37,10	23,00	27,62	51,80	22,00	29,44
15/05/2012	32,40	23,40	26,94	30,90	24,50	26,98	36,00	22,60	27,20	48,60	21,40	28,76
16/05/2012	33,70	23,60	27,92	32,40	25,00	28,40	37,30	22,80	28,30	50,60	21,70	30,50
17/05/2012	33,00	24,30	27,11	32,00	25,70	27,57	36,60	23,50	27,23	50,00	22,40	28,76
18/05/2012	33,20	22,80	27,04	31,70	24,30	27,25	36,60	21,80	27,25	50,70	20,50	28,81
19/05/2012	33,80	23,10	27,50	32,30	24,30	27,77	37,50	22,40	27,87	50,40	21,40	29,78
20/05/2012	33,40	24,00	28,32	32,30	25,40	28,72	36,80	23,20	28,79	49,70	22,20	30,78
21/05/2012	33,80	22,70	27,79	33,10	24,40	28,32	36,60	21,60	28,16	49,80	20,40	30,33
22/05/2012	32,10	23,20	27,08	30,40	24,10	26,85	35,80	22,40	27,33	51,30	21,20	28,64
23/05/2012	32,80	23,50	27,36	31,50	24,70	27,51	36,20	22,50	27,70	48,80	21,40	29,53
24/05/2012	33,10	23,90	27,14	31,10	24,50	26,99	37,10	23,20	27,59	50,40	22,20	29,30
25/05/2012	33,90	23,70	27,83	32,10	24,40	27,63	37,20	22,90	28,38	49,80	21,90	30,27
26/05/2012	33,20	23,60	27,55	32,10	25,10	27,84	37,20	22,60	27,90	52,30	21,30	29,90
27/05/2012	33,80	22,90	27,98	32,10	24,00	28,00	37,70	21,90	28,50	49,60	20,70	30,67
28/05/2012	34,50	23,90	27,79	33,30	25,40	28,29	37,10	23,30	28,20	53,30	22,50	30,36
29/05/2012	34,00	23,50	27,92	32,70	24,60	28,23	37,40	22,90	28,49	49,90	22,00	30,72
30/05/2012	33,20	24,40	28,03	32,20	25,70	28,48	36,70	23,70	28,44	48,20	23,00	30,27
31/05/2012	33,80	23,30	28,03	32,50	24,80	28,23	35,80	23,60	27,42	47,80	22,50	28,55
01/06/2012	32,50	24,10	27,25	31,30	24,80	27,43	35,80	23,60	27,42	47,80	22,50	28,55
02/06/2012	31,70	23,20	26,26	29,50	23,90	26,16	35,50	22,30	26,43	48,80	21,20	27,33
03/06/2012	31,10	23,60	25,59	29,60	24,30	25,73	34,10	23,00	25,75	42,90	22,10	26,64
04/06/2012	31,30	23,30	26,71	29,10	23,70	26,27	35,00	22,50	27,31	46,60	21,60	28,85
05/06/2012	33,20	22,80	26,61	32,10	24,00	27,20	35,20	21,90	26,42	41,70	20,80	26,62
06/06/2012	34,50	23,40	28,41	33,60	24,90	28,80	37,70	22,20	28,88	50,60	20,90	31,35
07/06/2012	34,60	24,50	28,37	33,20	25,60	28,53	38,50	23,90	28,93	50,80	23,10	31,23
08/06/2012	34,60	24,10	28,35	33,00	25,00	28,37	38,50	23,30	28,91	52,00	22,20	31,32
09/06/2012	33,70	24,60	27,73	32,30	25,70	27,88	36,90	23,70	27,96	49,50	22,70	29,33
10/06/2012	33,30	24,40	27,67	32,20	25,20	27,64	37,10	23,90	28,00	49,50	23,20	29,65
11/06/2012	33,70	23,90	28,26	32,80	25,00	28,56	37,20	23,20	28,79	50,10	22,10	30,86
12/06/2012	33,90	24,70	28,10	32,90	26,20	28,63	37,70	24,00	28,46	50,70	23,20	30,44
13/06/2012	32,30	23,70	26,77	31,10	24,50	27,12	35,70	22,90	26,90	48,40	21,90	28,00
14/06/2012	33,70	22,30	27,33	32,10	23,30	27,43	37,10	21,40	27,90	50,40	20,30	30,32
15/06/2012	33,90	23,50	27,39	31,80	24,60	27,56	37,70	22,80	27,60	50,90	21,60	28,98

16/06/2012	33,50	24,10	27,58	31,90	25,30	27,72	36,10	23,40	27,77	44,20	22,40	28,95
17/06/2012	33,90	23,20	27,40	32,00	24,40	27,60	37,50	22,20	27,61	48,20	21,00	28,78
18/06/2012	34,20	23,40	27,62	32,50	24,90	27,89	38,10	22,50	27,71	48,90	21,40	28,87
19/06/2012	33,70	23,50	27,40	32,00	24,90	27,63	36,70	22,80	27,41	46,50	21,70	28,20
20/06/2012	33,10	23,70	28,18	32,00	25,00	28,45	35,80	23,00	28,50	46,80	22,20	29,91
21/06/2012	34,80	23,50	27,76	33,20	25,30	28,22	38,10	22,60	26,92	49,10	21,40	28,42
22/06/2012							35,90	22,30	26,36	46,30	21,30	27,08
23/06/2012							36,50	21,80	27,40	44,70	20,70	28,34
24/06/2012							36,60	22,40	27,65	48,10	21,10	28,90
25/06/2012							34,80	22,20	26,41	42,80	21,20	27,24
26/06/2012							35,10	21,30	25,18	46,00	20,10	26,00
27/06/2012							36,50	21,80	27,83	47,90	20,80	29,28
28/06/2012							38,50	23,00	28,34	49,00	21,90	30,10
29/06/2012							37,30	22,70	27,66	47,40	21,60	29,09
30/06/2012							35,90	23,00	26,71	43,90	22,00	27,57
01/07/2012							36,30	22,90	27,10	45,30	22,10	27,94
02/07/2012							36,10	22,60	26,49	46,90	22,00	27,45
03/07/2012							35,50	22,30	27,01	45,40	21,30	28,15
04/07/2012							35,80	22,90	26,43	45,60	22,10	27,21
05/07/2012							36,60	22,90	27,70	45,20	22,20	28,87
06/07/2012							36,90	23,50	27,98	45,90	22,70	29,04
07/07/2012							37,70	23,50	28,80	48,90	22,50	30,46
08/07/2012							36,80	23,40	28,08	46,50	22,60	29,83
09/07/2012							33,90	23,50	26,22	41,00	22,90	26,92
10/07/2012	33,20	25,40	28,31	30,40	25,40	27,52	37,00	23,00	27,21	45,70	22,20	28,23
11/07/2012	31,90	23,80	27,00	30,20	24,60	26,70	34,80	23,00	27,10	45,60	22,00	27,71
12/07/2012	33,80	23,30	27,88	32,00	24,40	28,02	37,10	22,50	28,24	47,10	21,50	29,66
13/07/2012	33,80	24,20	27,80	32,30	25,20	28,26	37,00	23,50	27,99	49,70	22,50	29,29
14/07/2012	34,30	23,50	27,16	32,70	24,40	27,11	37,60	22,80	27,31	46,30	21,90	28,27
15/07/2012	33,20	24,10	27,66	31,90	25,20	27,83	36,30	23,60	27,95	45,10	23,00	29,10
16/07/2012	34,50	23,90	28,08	32,90	25,00	28,29	37,80	23,30	28,47	47,20	22,20	30,10
17/07/2012	34,40	24,00	28,45	33,20	25,50	28,89	37,40	23,20	28,92	49,70	22,20	31,11
18/07/2012	32,10	23,60	27,31	31,10	25,50	28,01	34,60	22,60	27,51	41,70	21,30	28,55
19/07/2012	34,20	22,90	27,41	32,90	24,30	27,75	37,70	22,00	27,74	46,50	20,80	29,21
20/07/2012	33,90	23,90	27,64	32,50	25,00	27,67	36,90	23,40	27,88	46,20	22,30	28,84
21/07/2012	32,30	24,30	27,63	31,10	25,20	27,89	35,60	23,30	27,86	43,90	22,90	28,80
22/07/2012	34,00	23,00	27,57	32,90	24,60	27,87	36,60	22,00	27,84	47,20	20,70	29,08
23/07/2012	33,90	23,30	28,28	32,50	24,60	28,39	36,90	22,30	28,76	46,80	21,10	30,23
24/07/2012	34,00	25,90	28,16	32,80	26,70	28,56	36,70	25,50	28,52	43,60	24,40	29,69
25/07/2012	34,40	24,40	27,76	32,90	25,40	28,28	37,80	23,50	28,02	47,80	22,50	29,30
26/07/2012	34,30	23,60	28,30	33,10	25,10	28,70	37,10	22,90	28,76	47,20	21,80	30,28
27/07/2012	34,50	24,70	28,97	33,70	26,10	29,43	37,20	24,10	29,47	47,80	23,30	31,09
28/07/2012	34,00	24,50	28,36	32,90	25,70	28,83	37,40	24,00	28,80	47,20	23,30	30,25
29/07/2012	34,70	24,00	28,57	33,70	25,20	29,01	37,70	23,30	28,97	47,70	22,30	30,54
30/07/2012	34,80	23,90	28,71	33,90	25,20	29,23	38,40	23,20	29,16	48,50	22,20	30,95
31/07/2012	34,70	24,00	29,69	33,90	25,20	29,92	37,60	23,50	30,76	48,80	22,70	33,95
01/08/2012	34,70	26,50	30,93	33,90	27,60	31,29	37,60	26,00	32,07	48,80	25,10	35,12
02/08/2012	34,90	23,50	29,06	34,20	24,90	29,59	38,30	22,70	29,57	50,50	21,40	31,61
03/08/2012	34,60	24,50	29,17	34,40	25,20	29,48	37,20	24,00	29,72	48,70	23,20	31,95
04/08/2012	34,50	24,90	27,90	32,20	24,50	27,43	38,10	24,50	28,47	49,60	23,80	30,35
05/08/2012	32,50	24,20	27,57	31,50	24,80	27,85	35,10	23,60	27,97	45,50	22,60	29,43
06/08/2012	34,10	24,50	28,82	33,20	25,30	29,07	36,70	24,10	29,38	49,50	23,20	31,20
07/08/2012	33,50	24,30	27,57	32,50	25,30	27,77	36,60	23,80	27,88	49,90	23,00	28,98
08/08/2012	34,00	23,80	26,37	32,20	24,40	26,32	37,60	23,10	26,66	48,90	22,10	27,76
09/08/2012	33,10	23,30	27,38	31,10	23,70	27,03	36,40	22,60	28,00	48,00	21,40	29,72
10/08/2012	34,10	23,90	27,90	32,60	24,90	28,07	37,40	23,60	28,43	49,30	22,80	30,18
11/08/2012	33,20	24,70	27,51	31,80	25,40	27,75	36,50	24,20	27,63	48,50	23,40	28,43
12/08/2012	33,00	23,90	27,62	31,70	24,60	27,56	36,30	23,40	27,97	46,80	22,50	29,19
13/08/2012	35,10	23,00	28,34	33,80	24,20	28,60	38,40	22,30	28,95	48,80	21,20	31,07
14/08/2012	33,60	24,10	27,55	32,20	24,60	27,73	37,30	23,40	27,91	50,10	22,50	29,26
15/08/2012	32,80	24,00	26,97	30,50	24,20	26,52	36,90	23,40	27,36	47,90	22,50	28,50
16/08/2012	33,40	23,40	27,88	32,20	24,40	28,03	36,40	22,60	28,43	47,30	21,50	30,30
17/08/2012	34,50	24,20	28,17	33,10	25,10	28,54	37,30	23,60	28,68	47,60	22,70	30,56
18/08/2012	34,60	23,90	28,54	33,60	24,90	28,86	37,70	23,30	29,10	49,80	22,40	31,22
19/08/2012	34,40	24,20	29,10	33,90	25,40	29,58	38,10	23,40	29,76	51,10	22,20	32,09
20/08/2012	33,80	24,50	28,61	32,80	25,60	29,06	37,60	23,70	29,18	48,40	22,60	31,05

21/08/2012	33,10	24,40	28,30	32,50	25,40	28,68	36,60	23,90	28,83	46,80	22,90	30,52
22/08/2012	34,00	24,70	28,02	33,00	25,40	28,39	37,40	24,10	28,33	48,50	23,30	29,69
23/08/2012	34,20	23,70	28,21	33,80	24,50	28,38	37,60	23,00	28,74	49,50	21,90	30,60
24/08/2012	34,10	23,50	28,26	33,10	24,10	28,25	37,70	22,60	28,92	49,90	22,00	30,89
25/08/2012	34,40	24,10	28,62	33,70	24,90	28,85	37,70	23,40	29,30	50,40	22,50	31,44
26/08/2012	34,90	24,10	28,84	34,20	24,70	29,12	38,40	23,60	29,53	49,90	22,60	31,70
27/08/2012	34,70	24,00	28,82	34,00	24,90	29,01	37,70	23,40	29,38	49,60	22,70	31,25
28/08/2012	33,50	23,80	27,09	31,90	24,80	27,44	37,00	23,00	27,21	50,20	21,80	28,25
29/08/2012	32,90	23,30	26,50	31,80	23,70	26,49	36,80	22,40	26,73	49,30	21,20	28,02
30/08/2012	33,40	23,80	27,78	31,20	23,90	27,34	37,30	23,30	28,42	49,50	22,40	30,06
31/08/2012	32,90	24,00	27,52	31,80	24,80	27,83	35,80	22,50	27,73	45,90	22,50	29,00
01/09/2012	32,50	23,80	26,87	31,50	24,60	27,35	36,00	23,10	26,97	47,50	22,10	27,96
02/09/2012	32,50	22,70	26,91	31,60	23,60	27,40	35,60	21,90	27,13	47,40	20,60	28,76
03/09/2012	33,90	23,80	27,24	32,40	24,50	27,33	37,30	22,30	27,61	50,00	22,50	29,52
04/09/2012	34,10	23,40	27,93	32,70	24,20	27,96	37,20	20,90	28,38	50,20	22,00	30,53
05/09/2012	33,50	23,50	28,13	32,30	24,60	28,37	37,20	22,80	28,49	49,00	22,00	29,99
06/09/2012	33,60	24,00	28,09	32,50	24,90	28,32	36,90	23,30	28,52	49,10	22,70	30,28
07/09/2012	33,50	23,90	27,91	32,70	24,90	28,21	36,50	22,10	28,16	47,70	22,50	29,73
08/09/2012	33,60	23,60	28,19	32,80	24,30	28,35	37,30	22,70	28,55	50,50	22,00	30,41
09/09/2012	34,40	24,10	28,39	33,40	25,00	28,65	38,30	23,40	28,87	51,60	22,50	30,81
10/09/2012	33,20	24,20	28,00	32,10	24,90	28,24	36,40	22,20	28,24	47,60	22,30	29,88
11/09/2012	32,90	23,80	26,72	31,90	24,10	26,71	36,60	23,10	27,19	48,60	21,90	28,32
12/09/2012	34,60	22,80	28,06	33,10	23,10	27,68	37,60	22,10	28,72	48,30	21,40	30,58
13/09/2012	35,40	23,60	28,86	34,30	24,50	28,93	38,70	23,20	29,49	50,10	22,20	31,49
14/09/2012	33,40	24,80	27,88	32,50	25,10	27,56	35,90	21,70	27,95	43,80	23,30	28,88
15/09/2012	34,30	24,00	27,96	31,90	24,20	27,31	38,20	22,20	28,76	47,90	22,70	30,96
16/09/2012	33,30	24,20	27,04	31,80	24,80	26,92	36,10	23,30	27,26	45,10	22,90	28,20
17/09/2012	33,00	23,60	28,00	31,70	24,10	27,80	35,30	22,80	28,26	45,30	22,20	29,96
18/09/2012	35,10	24,20	28,73	33,10	25,30	28,76	36,90	23,20	28,96	46,70	22,70	30,15
19/09/2012	33,20	23,80	28,18	32,00	24,90	28,37	36,40	23,30	28,55	46,90	22,40	29,89
20/09/2012	32,80	23,30	27,42	31,80	24,50	27,59	35,70	22,30	27,42	46,20	21,10	27,87
21/09/2012	32,90	23,90	28,12	31,70	24,80	28,20	35,10	23,20	28,37	45,90	22,20	29,84
22/09/2012	34,10	23,70	27,69	32,40	24,80	27,50	37,10	22,90	27,93	47,00	21,90	28,92
23/09/2012	33,00	23,90	27,45	31,50	24,70	27,21	35,70	23,20	27,65	45,00	22,20	28,95
24/09/2012	32,80	23,90	27,26	30,90	24,60	26,95	35,80	22,30	27,51	44,20	22,10	28,61
25/09/2012	32,20	24,20	27,71	30,40	24,90	27,60	37,40	23,70	28,10	44,50	23,00	29,01
26/09/2012	34,70	23,50	28,15	33,00	24,50	28,35	38,70	22,70	28,72	50,70	21,60	30,59
27/09/2012	32,80	23,40	26,37	31,30	23,40	26,33	35,70	22,80	26,45	48,70	21,80	27,66
28/09/2012	31,30	23,70	26,89	28,60	23,70	25,83	34,40	23,20	27,30	45,10	22,40	28,35
29/09/2012	32,70	23,70	27,40	31,20	24,30	27,28	35,00	23,10	27,98	44,80	22,30	29,21
30/09/2012	34,20	23,60	27,21	32,50	24,70	27,21	37,70	22,70	27,58	49,60	21,60	29,01
01/10/2012	33,10	23,70	27,80	31,50	24,30	27,54	35,60	23,20	28,47	48,20	22,30	30,37
02/10/2012	33,60	24,40	28,24	32,40	25,50	28,31	36,30	23,90	28,67	47,20	22,90	30,23
03/10/2012	33,60	24,60	27,95	32,10	25,30	27,98	36,20	24,20	28,33	46,70	23,30	29,54
04/10/2012	33,60	23,50	27,58	32,20	24,60	27,59	36,70	22,80	27,89	46,40	21,60	29,11
05/10/2012	32,20	23,80	27,66	31,00	24,70	27,74	34,90	23,10	27,87	44,90	21,90	28,74
06/10/2012	32,60	24,30	27,38	31,30	25,10	27,51	35,50	23,80	27,47	46,30	23,20	28,21
07/10/2012	34,10	24,00	27,70	32,60	24,50	27,81	37,10	23,40	28,05	47,50	22,60	29,46
08/10/2012	33,30	23,50	27,78	32,00	24,50	27,97	36,10	22,60	28,21	47,70	21,30	29,70
09/10/2012	32,80	24,70	28,32	31,90	25,70	28,47	35,70	24,10	28,74	46,70	23,20	30,08
10/10/2012	32,60	24,50	27,34	31,50	25,00	27,16	35,00	23,90	27,52	44,60	22,80	28,30
11/10/2012	32,80	23,70	27,62	31,30	24,20	27,22	35,20	23,00	28,18	44,50	21,80	29,66
12/10/2012	34,30	23,20	27,82	32,80	24,30	27,85	37,80	22,20	28,25	48,40	20,90	29,71
13/10/2012	31,70	24,00	26,83	31,10	24,90	27,53	33,00	23,50	26,53	36,20	22,70	26,02
14/10/2012	31,80	24,00	27,75	31,00	25,10	27,90	34,10	23,30	28,06	43,50	22,20	29,12
15/10/2012	33,20	25,00	28,24	32,20	25,70	28,35	35,50	24,40	28,62	44,70	23,70	29,86
16/10/2012	32,80	23,60	27,95	31,80	24,60	28,08	35,70	22,80	28,37	45,30	21,70	29,81
17/10/2012	32,90	23,90	27,93	31,70	24,90	28,05	35,90	23,30	28,18	45,90	22,30	29,15
18/10/2012	34,20	23,50	28,32	32,90	24,50	28,46	37,40	22,80	28,85	48,80	21,90	30,47
19/10/2012	33,10	24,40	28,30	32,50	25,10	28,52	35,70	23,80	28,72	46,40	22,80	30,19
20/10/2012	33,00	23,90	27,84	32,20	24,70	28,06	35,60	23,00	28,12	48,50	22,10	29,23
21/10/2012	34,60	23,30	28,20	33,70	24,10	28,44	37,40	22,40	28,77	50,80	21,20	31,02
22/10/2012	34,20	24,00	28,49	33,40	24,80	28,72	36,70	23,40	29,01	49,70	22,50	30,85
23/10/2012	34,10	24,40	28,80	33,50	25,10	28,90	36,50	23,90	29,28	47,00	23,00	30,78
24/10/2012	34,70	22,90	28,41	34,10	24,00	28,71	37,90	21,90	29,04	50,10	20,70	31,30
25/10/2012	33,00	24,00	28,34	32,30	25,00	28,57	35,20	23,30	28,68	42,90	22,30	29,84

26/10/2012	31,20	23,70	27,50	30,60	24,70	27,81	33,50	22,80	27,60	40,80	21,90	28,27
27/10/2012	32,60	23,60	27,42	31,70	24,30	27,79	34,90	22,80	27,62	44,40	21,80	28,59
28/10/2012	33,90	23,20	27,10	33,10	24,00	27,29	36,80	22,50	27,43	48,60	21,20	28,98
29/10/2012	33,70	23,50	27,95	32,50	24,10	27,78	36,40	22,80	28,46	49,00	21,80	30,27
30/10/2012	34,80	23,90	28,66	33,80	24,60	28,66	37,80	23,30	29,37	50,30	22,60	31,68
31/10/2012	34,30	23,60	28,53	33,60	24,50	28,76	38,20	22,70	29,21	51,40	21,40	31,48
01/11/2012	36,40	24,50	29,21	35,40	25,30	29,33	39,10	23,90	29,59	50,20	23,20	31,77
02/11/2012	34,00	24,50	28,84	33,40	25,30	28,97	37,00	23,90	29,38	46,90	23,00	30,91
03/11/2012	35,80	23,80	29,03	34,80	24,70	29,25	39,10	23,00	29,24	51,80	21,90	31,11
04/11/2012	35,10	23,60	28,82	34,50	24,60	29,11	38,40	22,80	29,44	52,20	21,60	31,36
05/11/2012	35,20	23,50	28,72	34,50	24,50	28,51	38,60	22,50	29,32	50,90	21,30	31,10
06/11/2012	33,30	25,00	27,93	31,50	25,30	27,47	37,70	24,30	28,34	51,10	23,20	29,55
07/11/2012	33,10	23,80	27,62	31,60	24,50	27,39	36,10	23,00	27,91	47,30	21,90	28,98
08/11/2012	32,10	23,60	27,69	30,70	24,60	27,68	35,50	22,80	28,00	49,20	21,80	29,04
09/11/2012	33,80	23,00	27,83	32,10	24,00	27,99	37,30	22,00	28,19	49,20	21,00	29,70
10/11/2012	33,20	23,40	28,17	31,90	24,60	28,11	36,60	22,50	29,01	49,20	21,30	31,56
11/11/2012	34,10	26,40	30,14	32,80	27,60	30,30	36,80	25,70	30,62	44,50	24,90	31,79
12/11/2012	35,00	24,00	28,65	33,10	25,30	28,89	38,60	23,20	29,24	51,40	22,20	31,27
13/11/2012	33,30	23,90	28,27	32,20	25,20	28,54	36,30	23,00	28,60	44,90	21,70	29,86
14/11/2012	32,70	24,70	27,67	31,60	25,60	27,64	36,40	24,10	27,83	48,20	23,30	28,68
15/11/2012	33,70	23,40	27,97	32,40	24,30	28,02	36,90	22,60	28,53	49,40	21,80	30,35
16/11/2012	34,70	24,30	28,51	33,30	25,30	28,67	37,20	23,50	29,06	49,00	22,80	31,11
17/11/2012	34,40	23,70	28,40	32,70	25,10	28,56	37,10	22,90	28,79	47,60	21,60	30,06
18/11/2012	35,20	23,50	28,92	33,80	24,80	28,98	38,70	22,50	29,51	52,00	21,20	31,41
19/11/2012	34,00	24,40	28,64	32,30	25,60	28,72	36,80	23,80	28,90	44,80	22,70	29,93
20/11/2012	33,70	25,20	28,90	32,40	26,00	28,89	36,80	24,50	29,28	46,30	23,40	30,45
21/11/2012	34,20	24,10	28,62	32,90	25,30	28,68	36,70	23,30	28,93	44,60	22,20	29,94
22/11/2012	35,10	24,40	28,92	33,20	25,30	28,94	39,00	23,90	29,39	52,50	23,10	30,74
23/11/2012	35,10	23,40	29,00	33,70	24,80	29,09	38,30	22,40	29,55	51,90	21,10	31,32
24/11/2012	33,00	25,10	28,47	31,50	26,20	28,50	35,60	24,40	28,65	43,80	23,50	29,25
25/11/2012	34,00	24,90	28,46	32,40	26,00	28,38	37,20	24,20	28,65	46,20	23,40	29,68
26/11/2012	34,80	23,60	28,69	33,00	24,80	28,73	37,00	22,80	29,09	48,20	21,60	30,77
27/11/2012	35,80	25,80	30,06	34,60	26,40	29,93	39,10	23,60	30,67	50,40	24,60	33,19
28/11/2012	34,70	24,90	29,30	33,50	26,10	29,40	37,50	22,70	29,64	47,10	23,10	31,61
29/11/2012	33,60	25,70	27,81	31,90	25,70	27,58	36,80	25,10	27,93	45,80	24,40	28,60
30/11/2012	30,30	23,70	26,26	28,80	24,00	26,03	32,20	23,30	26,46	37,50	22,50	27,05
01/12/2012	33,80	23,50	27,54	32,10	24,30	27,20	36,50	22,90	27,85	46,30	21,80	29,33
02/12/2012	34,30	24,90	28,01	32,10	25,20	27,87	38,00	24,40	28,43	50,00	23,50	29,87
03/12/2012	35,10	24,30	28,00	33,30	25,50	27,98	38,10	23,10	28,48	47,70	22,90	30,39
04/12/2012	34,10	23,90	28,04	32,10	24,80	27,99	37,40	23,40	28,38	48,80	22,50	29,81
05/12/2012	34,80	23,70	28,67	33,10	24,80	28,75	38,30	22,90	29,21	50,50	21,90	31,07
06/12/2012	33,30	24,70	27,83	31,60	25,80	27,81	36,10	23,60	27,78	52,20	23,20	28,65
07/12/2012	32,70	25,40	28,29	31,10	25,70	28,14	35,30	22,10	28,50	49,80	23,90	29,99
08/12/2012	34,40	24,60	28,76	32,70	25,70	28,61	37,70	23,20	29,12	49,30	23,10	30,98
09/12/2012	29,90	24,20	26,07	28,70	25,40	26,54	31,60	23,40	26,02	37,40	22,40	26,33

Appendix B - Climatological data for the entire period studied

Date	Temperature (°C)		Humidity (%)	Pressure (mbar)	Wind (m/s)	Sunshine	Rain
	Bulb Dry	Wet bulb	Inst.	Inst.	Vel.	h/day	(mm/24h)
1/12/2011	28,66667	25,86667	79,33333	1009,333	1,066667	3,5	5,6
2/12/2011	28,2	26	83	1008,467	1,4	7	11,2
3/12/2011	28,53333	25,23333	75,66667	1008,033	1,533333	8,2	0,5
4/12/2011	29,03333	25,26667	72,33333	1007,767	1,066667	10,3	0
5/12/2011	28,83333	25,96667	79,33333	1006,967	1,2	9,7	0
6/12/2011	29,3	25,76667	74	1006	1,6	9,4	11,2
7/12/2011	28,73333	26,06667	80,33333	1006,833	1,333333	7,7	2,6
8/12/2011	28,1	25,93333	84,33333	1008,033	1,066667	5,4	5,5
9/12/2011	28,6	26,1	80,66667	1008,2	1,266667	10,1	36,2
10/12/2011	28,7	26	80,33333	1008,333	1,033333	9,4	0
11/12/2011	28,13333	25,63333	81,33333	1007,9	0,966667	8,2	20,6
12/12/2011	28,56667	26,6	85,33333	1007,967	1,366667	8,1	0,1
13/12/2011	28,33333	25,86667	81,33333	1007,633	1,966667	8,9	20,6
14/12/2011	28,73333	26,23333	81,66667	1007,133	1,3	7,6	4
15/12/2011	28,13333	25,73333	81	1008	1,466667	6,4	0
16/12/2011	28,7	25,1	74,66667	1007,933	1,566667	6,9	7,6

17/12/2011	28,96667	25,63333	75,33333	1008,667	2,033333	8,5	0
18/12/2011	28,53333	26,03333	80,33333	1008,9	1,066667	7,7	4,4
19/12/2011	28,83333	25,23333	73,33333	1009	1,6	7,5	6,6
20/12/2011	29,23333	26,06667	78	1009	0,7	8,7	0
21/12/2011	27	25,16667	85	1008,933	0,833333	5,3	3
22/12/2011	28,86667	25,63333	76	1007,833	2,1	7,9	19,6
23/12/2011	28,53333	25,83333	79	1008,067	1,033333	4,7	0
24/12/2011	29,46667	26,1	76,33333	1009,533	1,633333	7,3	0,2
25/12/2011	27,36667	25,2	84	1010,1	1,6	6,1	0
26/12/2011	28,2	26,26667	85	1010,167	1,266667	5,8	20
27/12/2011	27,26667	25,76667	88,66667	1011,367	0,666667	2,3	1,3
28/12/2011	27,83333	25,86667	85	1011,633	0,8	3,4	17,8
29/12/2011	28,13333	25,7	80,66667	1010,2	1,433333	7	1,1
30/12/2011	29,13333	26,26667	79,66667	1009,3	1,233333	9,6	2
31/12/2011	28,53333	26,26667	82	1009,367	1,6	7,6	0
1/1/2012	28,16667	25,7	80	1010,067	0,533333	5,6	0,4
2/1/2012	28,6	25,13333	74,33333	1008,867	1,866667	8,1	0,8
3/1/2012	29,46667	25,13333	68,33333	1009,267	1,866667	5,9	0
4/1/2012	27,96667	26,16667	87,33333	1009,733	1,1	3,5	0
5/1/2012	26,66667	25,13333	87,66667	1010,067	0,466667	0	13,4
6/1/2012	26,13333	25,26667	92,66667	1009,367	1,066667	4,8	32,4
7/1/2012	28,53333	25,86667	79,66667	1008,867	1,2	9,4	34,8
8/1/2012	28,4	26,16667	82,66667	1009,4	1,1	8,6	3,2
9/1/2012	27,43333	26,16667	89,66667	1008,967	1,5	6,5	4,4
10/1/2012	26,73333	25,13333	87	1008,3	1,1	5,4	13,7
11/1/2012	26,93333	25,56667	88,66667	1008,967	1,033333	3,5	29
12/1/2012	26,73333	25,3	87,33333	1009,3	1,433333	6,9	43,4
13/1/2012	27,4	25,43333	86	1009,367	1,6	7,7	23,5
14/1/2012	28,2	26,3	85,33333	1009,067	0,5	6,8	6,3
15/1/2012	27,86667	26,03333	86,33333	1009,833	1,333333	2,2	0,4
16/1/2012	22,9	22,4	95,66667	1010,567	0,866667	0	5
17/1/2012	26,63333	25,43333	89,33333	1009,067	0,266667	0,4	24,2
18/1/2012	25,96667	25,03333	92,33333	1008,933	0,266667	3,3	0
19/1/2012	23,63333	23,16667	95	1009,433	0,666667	3,2	15,1
20/1/2012	27	25,23333	86	1008,633	0,566667	6,7	42,4
21/1/2012	26,2	25,1	90,66667	1008,433	1,2	1,3	0,1
22/1/2012	25,7	25,2	94,33333	1008,4	0	2,9	38,7
23/1/2012	24,83333	24,03333	93	1008,867	0,833333	2,8	9,2
24/1/2012	25,93333	24,46667	88	1008,233	1,566667	5,3	32,7
25/1/2012	27,66667	25,96667	86,33333	1008,4	1,066667	8,6	34,1
26/1/2012	28,16667	26,2	85,66667	1008,1	0,6	6,9	4,2
27/1/2012	28,23333	26,53333	86,66667	1007,867	0,866667	7,3	8,7
28/1/2012	28,36667	26,46667	85	1009	1,133333	9,6	24,2
29/1/2012	27,73333	25,53333	82,66667	1008,433	1,133333	8	3,7
30/1/2012	28,53333	26	81	1007,867	1,466667	7	3,2
31/1/2012	28,6	25,9	79,33333	1007,867	0,866667	8,1	0
1/2/2012	28,93333	25,73333	77	1007,833	0,966667	9,3	3
2/2/2012	29,03333	25,46667	74,33333	1007,167	2,233333	6,6	1,8
3/2/2012	27,9	25,2	80	1007,267	0,866667	9	1
4/2/2012	27,66667	24,93333	80,33333	1007,933	1,633333	4,4	5,8
5/2/2012	28,4	25,8	81,33333	1008,033	0,866667	7,8	21,3
6/2/2012	28,26667	25,7	81	1008,967	0,866667	5,3	4,4
7/2/2012	26,86667	25,6	89,33333	1008,3	0,9	6	8,4
8/2/2012	26,13333	25	91	1008,133	1,3	4,1	44,7
9/2/2012	26,43333	25,2	89,66667	1009,367	0,733333	0	13,8
10/2/2012	26,1	25,26667	93	1009,667	0,333333	0	9,1
11/2/2012	24,3	23,63333	93,66667	1010	0,866667	0,7	0,9
12/2/2012	25,76667	24,4	89	1009,1	0,7	4,1	18,2
13/2/2012	25,76667	24,36667	89	1008,5	0,266667	2,8	3,7
14/2/2012	24,53333	24,06667	95	1009,3	0,333333	2	2,4
15/2/2012	25,1	24,53333	95	1010,4	0	1,2	29,9
16/2/2012	26,73333	25,13333	88	1009,833	0,266667	4,1	2,7
17/2/2012	23,86667	23,4	95	1010,9	0,6	1,2	12,9
18/2/2012	26,06667	25,2	93,33333	1008,633	0	3,9	31,7
19/2/2012	26,66667	25	87	1009,767	0,366667	5,4	1,2
20/2/2012	26,73333	25,33333	88,66667	1009,533	0,7	6,3	18,6
21/2/2012	27	25,73333	90,66667	1008,767	0,566667	5,3	13,2

22/2/2012	27,5	25,43333	84,33333	1008,1	1,7	10,5	20,5
23/2/2012	26,83333	25,76667	92,33333	1008,1	0,366667	2,8	31
24/2/2012	26,7	25,56667	90,66667	1007,833	0,333333	5,9	19,4
25/2/2012	25,73333	24,83333	93	1008,267	0,733333	0,8	30,1
26/2/2012	25,33333	24,83333	95,33333	1010	0,266667	0,2	1,4
27/2/2012	24,93333	24,5	95,66667	1009,633	0,3	1,1	15,8
28/2/2012	25,46667	25,1	96,33333	1009,267	0,166667	1,6	27,9
29/2/2012	26,46667	25,6	93	1009,033	0,133333	6,5	17,1
1/3/2012	26	25	91	1009,6	1,066667	3,3	13,6
2/3/2012	26,7	25,56667	91,66667	1009,8	0,433333	3,2	8,4
3/3/2012	26,06667	25,23333	92,66667	1010,267	0,7	4,2	14,8
4/3/2012	26,56667	25,46667	90,33333	1009,6	0,333333	7,8	35,3
5/3/2012	25,4	24,3	90,66667	1009,7	0,866667	5,2	2,4
6/3/2012	26,06667	24,66667	88	1008,7	0,3	4,9	11,3
7/3/2012	26,56667	24,83333	87,33333	1009,6	0,533333	1,9	11,8
8/3/2012	25,93333	24,73333	90	1009,167	1	4,7	0,2
9/3/2012	25,86667	25,03333	92,66667	1009,533	0,166667	5,5	42,4
10/3/2012	26,23333	25,2	91,33333	1010,133	0,933333	5,4	14,5
11/3/2012	26,6	25,86667	94,66667	1010,2	0,866667	5,1	12,2
12/3/2012	25,56667	25,06667	94,33333	1009,667	0,533333	2,1	81,9
13/3/2012	26,33333	25,4	91	1009,9	0,533333	3,7	49,6
14/3/2012	25,6	24,93333	94,66667	1010,333	0,266667	4	33,1
15/3/2012	26,96667	25,23333	86,66667	1010,7	0,6	2,9	32,8
16/3/2012	26,83333	25,6	89,66667	1010,233	1,133333	4,9	2,8
17/3/2012	29,36667	25,8	74,66667	1008,7	1,6	8,1	10,9
18/3/2012	26,96667	25,23333	86,33333	1008,433	1,033333	6,7	0,2
19/3/2012	25,93333	24,86667	90,66667	1008,833	0	1	35,8
20/3/2012	26,16667	25,03333	91	1009,233	0,233333	4	2,8
21/3/2012	25,86667	24,66667	89,33333	1008,667	0,633333	4,7	60,3
22/3/2012	27,23333	25,6	87	1007,833	0,633333	3,6	8,6
23/3/2012	27,23333	25,66667	89,33333	1008,6	0,533333	4,6	62,7
24/3/2012	27,33333	25,4	85,66667	1010	1,433333	5,6	32,4
25/3/2012	26,66667	25,06667	87,33333	1010	0,1	2,4	24,1
26/3/2012	26,06667	25,03333	91,33333	1008,967	1,033333	2,4	23,4
27/3/2012	27	25,4	87,66667	1009,867	2,066667	4,4	38
28/3/2012	24,63333	24,2	96	1011,1	0	0	49,2
29/3/2012	28,23333	25,86667	82,33333	1010,467	1,666667	8,9	10,3
30/3/2012	27,63333	25,96667	86,33333	1011,733	1,366667	6,1	6,2
31/3/2012	26,83333	25,56667	89,33333	1011,8	0,7	3,9	10,5
1/4/2012	28,43333	26,66667	85,33333	1009,533	0,7	5	22,3
2/4/2012	27,7	25,86667	85	1007,6	1,366667	8	7,9
3/4/2012	27,5	25,73333	87	1008,967	0,9	6,6	19,2
4/4/2012	28,1	25,7	82,66667	1008,6	1,333333	8,8	3,6
5/4/2012	28,6	25,76667	79	1007,967	1,533333	7	17,2
6/4/2012	29,3	26,36667	78,33333	1008,233	2,2	5,3	1,2
7/4/2012	27,63333	25,56667	83,33333	1008,3	2,266667	6,6	0
8/4/2012	27,06667	25,63333	88	1008,767	1,633333	6,2	12,8
9/4/2012	24,73333	23,76667	91	1009,067	0,166667	0	29,2
10/4/2012	26,96667	25,43333	88	1009,767	1,733333	7,4	0
11/4/2012	26,6	25,23333	88,33333	1010,9	2,766667	7,1	37,6
12/4/2012	26,6	25,4	89,66667	1009,267	0,533333	7,1	20,8
13/4/2012	28,16667	26,16667	85,66667	1009,533	0,966667	6,1	38
14/4/2012	26,06667	24,76667	88,66667	1010,333	1,566667	1,8	1,2
15/4/2012	26,03333	24,86667	90,33333	1010,367	0,933333	2,3	8,4
16/4/2012	26,53333	25,1	89,33333	1009,967	0,766667	1,1	15,3
17/4/2012	26,1	24,96667	90,33333	1011,067	0,766667	6,3	5,6
18/4/2012	27,9	25,86667	84,66667	1010,067	1,1	7,5	46,2
19/4/2012	28,7	26,43333	83	1009,1	1,066667	7,5	16,2
20/4/2012	27,23333	25,6	86,66667	1008,7	1,4	7,9	3,8
21/4/2012	26,96667	25,76667	89	1010,367	1,133333	3	16,9
22/4/2012	27,56667	25,8	85	1010,533	1,266667	6,7	14,8
23/4/2012	27,4	25,7	86,66667	1009,733	1,933333	5,2	3,1
24/4/2012	28,33333	25,83333	82	1008,733	1,1	7,9	17
25/4/2012	28,33333	26,23333	85	1009,133	0,766667	8,1	6,1
26/4/2012	27,5	25,86667	87	1009,5	0,833333	3,8	0
27/4/2012	28,8	26,36667	81,66667	1008,767	1,533333	8	2,9
28/4/2012	27,33333	25,66667	86,66667	1008,467	0,666667	5,1	0

29/4/2012	28,1	25,86667	83,33333	1008,633	0,8	7,8	15,4
30/4/2012	28,33333	26	82,66667	1010,133	0,366667	6,6	0,1
1/5/2012	28	25,43333	80	1010	0,933333	7,1	0
2/5/2012	28,3	25,63333	79,33333	1009,633	1,266667	8	0,7
3/5/2012	28,53333	26,5	85	1009,7	1,366667	6,4	0
4/5/2012	28	25,93333	84	1009,133	1,033333	6,9	6,6
5/5/2012	28,4	25,63333	78,66667	1009,533	0,766667	7,4	0,6
6/5/2012	28,96667	26,1	79	1009,933	0,8	7,7	6,6
7/5/2012	28,16667	25,76667	82	1010,3	0,7	7,9	0,9
8/5/2012	26,9	25,2	86	1010,533	1,766667	6,4	22,3
9/5/2012	28,83333	26,13333	79,66667	1009,667	0,533333	8,4	9,8
10/5/2012	28,46667	25,5	78	1009,2	1,1	7,6	5,7
11/5/2012	25,86667	24,66667	89,66667	1010,067	0,966667	5,4	17
12/5/2012	27,3	25,73333	87,66667	1009,433	0,9	7,3	10,3
13/5/2012	26,53333	24,83333	86,33333	1008,833	1,1	5,7	6,2
14/5/2012	28,5	25,93333	82	1009,267	1,4	7,6	0,1
15/5/2012	27,73333	25,63333	84,66667	1010,133	0,866667	6,2	32,2
16/5/2012	28,76667	25,93333	79,33333	1010,233	0,966667	8,7	8,7
17/5/2012	26,43333	25,03333	88,33333	1009,667	2,366667	5,7	2,8
18/5/2012	28,56667	25,46667	78,66667	1009,033	1,066667	8,7	7
19/5/2012	28,86667	25,5	75,33333	1009,2	2,166667	8,4	0
20/5/2012	28,6	25,4	75,66667	1010,3	2,1	9,3	0
21/5/2012	28,4	25,26667	76,66667	1011,167	1,533333	9,8	0
22/5/2012	28,53333	25,73333	78,33333	1010,8	1,033333	8,8	4,2
23/5/2012	28,33333	25,8	80,33333	1011,433	1,166667	9,1	0
24/5/2012	28	25,8	83	1011,267	0,966667	7,7	11,5
25/5/2012	28,76667	25,93333	79,66667	1011,133	1,466667	9,2	28,4
26/5/2012	28,2	25,43333	79,33333	1010,933	0,766667	9,4	4,8
27/5/2012	29,2	25,66667	75	1011,433	1,266667	9,6	20,3
28/5/2012	28,33333	25,4	78	1011,3	1,1	9,1	0
29/5/2012	28,83333	26,06667	79,33333	1010,267	1,466667	9,3	10,8
30/5/2012	27,9	25,5	82,66667	1010,767	1,033333	5,4	0
31/5/2012	28,9	25,9	79,33333	1011,9	1,6	10,4	7,6
1/6/2012	27,8	25,63333	84,33333	1013	1,3	7,6	1,4
2/6/2012	27,66667	25,63333	83,66667	1011,933	1,2	6,9	15,8
3/6/2012	25,76667	24,66667	90,66667	1011,933	0,466667	3,6	5,1
4/6/2012	27,7	25,23333	80,66667	1011,5	0,966667	7,5	21,3
5/6/2012	29,06667	25,6	75	1012	2	9,5	2,8
6/6/2012	28,93333	25,86667	77,66667	1011,4	1,266667	10,2	0
7/6/2012	28,36667	25,53333	79	1010,4	1,233333	9	4,8
8/6/2012	28,9	26,16667	80	1009,333	1,433333	10,1	56,6
9/6/2012	27,43333	25,6	85,66667	1010,5	1,133333	6,6	15
10/6/2012	29,06667	26,16667	78,33333	1010,467	0,566667	8,4	11,8
11/6/2012	29,1	26,2	78,66667	1009,967	1,3	10,1	4
12/6/2012	28,9	25,9	78,33333	1009,733	1,533333	9,3	0
13/6/2012	27,46667	25,5	85	1010,8	1,1	8,4	1
14/6/2012	28,26667	25,2	77,33333	1010,9	1,366667	10,3	3,1
15/6/2012	27,7	25,5	83,33333	1010,933	0,566667	7,8	11,4
16/6/2012	28,03333	25,96667	84	1011,733	0,833333	6,6	2,4
17/6/2012	28,36667	25,6	79,66667	1011,433	0,8	8,6	9,3
18/6/2012	27,96667	25,5	81,33333	1011,533	1,033333	8,6	2,6
19/6/2012	28,03333	25,9	84	1012,067	0,9	7,1	3,4
20/6/2012	28,3	25,83333	81,66667	1011,433	0,633333	9,6	14,2
21/6/2012	27,9	25,23333	80	1010,767	1,2	9,2	0,3
22/6/2012	27,53333	25,53333	84,66667	1012	0,966667	6,8	9,2
23/6/2012	28,53333	25,6	77,66667	1013,4	1,733333	9,4	36,5
24/6/2012	28,23333	25,33333	78,33333	1013,067	0,866667	9,1	0
25/6/2012	27	24,63333	82	1012,633	1,1	7,1	4,4
26/6/2012	25,06667	23,93333	90,66667	1012,533	0,666667	5	10,6
27/6/2012	28,13333	25,23333	79,66667	1012,733	1	9,9	30,2
28/6/2012	27,96667	25,23333	80	1012,933	1,533333	9	4,3
29/6/2012	27,7	24,96667	80	1011,533	1,3	8,8	37,4
30/6/2012	26,73333	25,06667	87	1011,267	0,9	5,6	1,1
1/7/2012	27,46667	24,86667	80,66667	1010,867	1,266667	8,4	1,1
2/7/2012	27,53333	25,26667	83,66667	1010,733	0,7	7,2	33
3/7/2012	27,93333	24,93333	78	1010,2	1,233333	8,3	3,4
4/7/2012	26,6	25,13333	88,33333	1009,5	0,2	5,4	4,6

5/7/2012	28,26667	25,5	79,66667	1009	1,066667	8,4	13,6
6/7/2012	28,46667	25,33333	77,33333	1009,167	1,466667	7,5	5,2
7/7/2012	28,66667	25,46667	76,33333	1010,467	1,466667	9,8	13,6
8/7/2012	27,76667	25,5	83,33333	1010,5	0,733333	7,3	12,8
9/7/2012	26,06667	25,03333	91	1010,033	0,7	4	75,3
10/7/2012	28,06667	25,8	83	1009,433	0,966667	8,9	19
11/7/2012	28,1	25,93333	82,66667	1008,5	2	6,4	14,9
12/7/2012	28,36667	25,83333	81,66667	1009,133	0,7	10	8,9
13/7/2012	28,5	25,73333	79,66667	1011	0,4	7,4	0
14/7/2012	28,53333	25,86667	80	1011,2	1,166667	7	27,1
15/7/2012	28,1	25,83333	82,33333	1011,3	1,266667	7,5	1,3
16/7/2012	28,43333	25,53333	79	1011,033	0,966667	9,6	16,4
17/7/2012	29,13333	25,76667	75,66667	1011,733	1,466667	9,7	20,3
18/7/2012	27,76667	25,4	82,66667	1013	1,233333	4,9	0
19/7/2012	28,46667	25,6	79	1012,1	1	9,2	1
20/7/2012	28,56667	25,8	79,33333	1011,467	1,866667	8,2	0,8
21/7/2012	26,76667	24,73333	83	1012,433	1,2	7,8	0,1
22/7/2012	28,03333	25,06667	78	1012,833	1,666667	9,9	2,4
23/7/2012	28,93333	25,7	76,66667	1012,333	1,7	10,1	16
24/7/2012	28,1	25,53333	81	1012,167	1,866667	5,1	0
25/7/2012	27,86667	25,2	79	1012,4	1,533333	8,3	1
26/7/2012	28,7	25,76667	79,66667	1011,933	1,2	9,6	8
27/7/2012	29,2	25,6	74,66667	1012,1	1,866667	9,2	0
28/7/2012	28,3	25	76,33333	1011,267	1,933333	8,7	0
29/7/2012	29,06667	25,96667	77	1009,9	1,3	8,8	13,4
30/7/2012	28,96667	25,76667	76,33333	1011,5	1,233333	9	0
31/7/2012	28,66667	25,6	77,66667	1011,367	1,133333	9,6	0
1/8/2012	29,03333	25,33333	73,66667	1011,8	2,033333	9,8	0
2/8/2012	29,5	25,56667	73	1011,067	1,033333	9,9	0
3/8/2012	29,7	25,93333	73	1010,867	1,5	8,8	0
4/8/2012	28,6	25,46667	77,66667	1011,267	0,533333	5,5	28,4
5/8/2012	28,03333	25,33333	80,66667	1011,8	0,9	4	12,2
6/8/2012	29,56667	26,36667	77,33333	1011,067	1,766667	9,6	0
7/8/2012	28,2	25,86667	82	1011,333	1,066667	5,7	0,7
8/8/2012	27	25,03333	84,66667	1011,467	1,766667	3,9	2,3
9/8/2012	28,26667	25,3	78	1011,033	1,966667	9	10
10/8/2012	28,36667	25,63333	79	1011	1,766667	9,2	0
11/8/2012	28,26667	25,23333	77,33333	1011,067	2,033333	8,4	0,1
12/8/2012	27,73333	25	79	1011,4	2,066667	9	0
13/8/2012	28,56667	25,16667	76	1012,2	1,6	9,8	0,4
14/8/2012	28,16667	25,66667	81	1011,367	1,166667	8,5	0
15/8/2012	27,8	25,3	81	1011,333	1,2	7,6	3,3
16/8/2012	28,13333	25,1	77,33333	1011,467	1,3	9,5	7,2
17/8/2012	28,46667	25,4	78,66667	1011,467	1,366667	8,7	0
18/8/2012	28,6	25,4	76,33333	1011,3	1,333333	9,4	7,1
19/8/2012	28,9	25,9	78	1011,667	0,933333	9,4	0
20/8/2012	28,6	25,53333	77,33333	1012,033	1,1	8,7	0
21/8/2012	28,13333	25,23333	78	1012,067	1,966667	9	0
22/8/2012	28,83333	25,26667	75	1011,567	1,433333	8,3	0
23/8/2012	27,93333	24,76667	78	1011,4	1,866667	9,9	0,1
24/8/2012	28,66667	25,1	74,33333	1010,633	2,133333	9,6	9,3
25/8/2012	28,7	24,66667	71,33333	1011,7	2,2	9,6	0
26/8/2012	29,1	25,06667	71	1012,533	2,1	10,3	0
27/8/2012	29,36667	25,7	74	1013,5	1,533333	10	0
28/8/2012	28,1	25,7	81,66667	1013,567	1,066667	7,9	0,2
29/8/2012	28	25,43333	81,33333	1012,567	1,2	8,1	0,6
30/8/2012	28,96667	25,86667	77,66667	1011,8	1,3	9,1	30,3
31/8/2012	28,46667	25,33333	77	1011,533	1,9	8,7	0
1/9/2012	26,6	24,9	86,66667	1012	0,9	7,3	9,6
2/9/2012	27,53333	25,5	85	1013,2	1,766667	9	9
3/9/2012	28,2	25,46667	80,33333	1013,633	0,6	8,1	14,2
4/9/2012	28,8	25,46667	75,33333	1012,833	1,966667	10	4,9
5/9/2012	28,86667	25,23333	74,33333	1012,7	2,166667	9,8	0
6/9/2012	28,96667	25,36667	74	1012,633	0,866667	10,3	0
7/9/2012	29,3	25,7	74,66667	1012,8	1,533333	8,7	0
8/9/2012	29,36667	25,46667	72,33333	1013	2	10,1	0
9/9/2012	28,9	25,66667	76,66667	1012,533	1,833333	8,9	0

10/9/2012	28,7	26,4	82,33333	1011,733	1,866667	6,8	0
11/9/2012	28,13333	25,4	79,66667	1011,1	0,966667	5,8	0
12/9/2012	28,83333	25,03333	72	1011,033	1,9	9,7	16,1
13/9/2012	29,7	25,2	69	1011,1	2,166667	9,8	0,3
14/9/2012	27,7	25,96667	87,33333	1011,433	1,566667	8,3	0
15/9/2012	28,73333	25,43333	76	1011,733	2,2	9,6	38
16/9/2012	28,4	25,7	79,33333	1011,733	1,766667	0,5	0
17/9/2012	28,66667	25,73333	77,66667	1011,8	1,3	7,4	0
18/9/2012	29,4	25,66667	72	1011,2	1,733333	6,7	0,4
19/9/2012	28,33333	25,4	77,66667	1011,2	1,7	7,8	0
20/9/2012	27,36667	25,23333	83,33333	1010	2,2	6,7	1,2
21/9/2012	29,3	25,96667	74	1009,4	1,8	9,5	7,6
22/9/2012	29,2	25,73333	75	1010,833	1,933333	7,6	0
23/9/2012	29,13333	25,9	77,33333	1009,667	1,366667	8,1	0,9
24/9/2012	28,5	25,4	77,33333	1007,333	1,333333	7,4	0,3
25/9/2012	27,96667	25,73333	54	1011,1	1,266667	7,8	0,3
26/9/2012	28,76667	25,4	75,33333	1011,767	1,433333	9,7	23,5
27/9/2012	27,4	25,1	83	1011,267	1,566667	7,5	0
28/9/2012	28	25,56667	81,33333	1012	1,366667	7,7	63,1
29/9/2012	27,93333	25,33333	81,33333	1012,733	0,866667	7	1,2
30/9/2012	28,26667	25,23333	77,33333	1010,6	1,7	8,3	6,4
1/10/2012	29,26667	25,83333	75,33333	1009,433	1,533333	9,9	7,2
2/10/2012	29,63333	26,36667	76,66667	1010,133	0,866667	8,8	0
3/10/2012	28,96667	25,36667	74,33333	1011,367	1,1	7,5	0
4/10/2012	28,83333	25,76667	77,66667	1010,167	1,7	8,5	0
5/10/2012	28,83333	26	79,33333	1008,933	1,433333	6,3	3,7
6/10/2012	28,53333	25,3	76,66667	1009,133	1,866667	6,7	0
7/10/2012	29,1	25,06667	72	1009,667	2	9,5	0
8/10/2012	28,8	25,73333	77,66667	1010,133	1,1	8,1	5,4
9/10/2012	28,13333	26,1	84	1009,7	0,433333	3,8	4,5
10/10/2012	26,66667	25,23333	89,33333	1009,167	1,366667	3,5	0,8
11/10/2012	29,06667	25,3	73	1010	1,6	8,3	11,6
12/10/2012	29,3	25,63333	74,33333	1009,867	1,133333	8,8	0
13/10/2012	29,03333	25,56667	75	1008,833	1,966667	6,7	0
14/10/2012	28,16667	24,93333	76	1009,133	1,933333	1,8	0
15/10/2012	28,93333	25,6	75,33333	1008,833	1,266667	3,9	0
16/10/2012	28,76667	25,9	79	1008,733	1,533333	2,4	0,2
17/10/2012	29,1	25,5	74,33333	1009,467	1,733333	3,9	0
18/10/2012	29,8	26	73,33333	1008,8	1,8	8,1	0,1
19/10/2012	29,46667	25,73333	73,33333	1008,433	1,966667	9,1	0
20/10/2012	28,7	25,66667	77,33333	1008,733	1,433333	6,8	0
21/10/2012	29,3	25,76667	75	1007,9	1,7	9,5	2,1
22/10/2012	29,7	25,8	71,66667	1007,033	1,5	10,4	0
23/10/2012	29,73333	25,46667	69,33333	1007,933	2,066667	9,3	0
24/10/2012	29,1	25,33333	73,66667	1007,733	1,4	10	0
25/10/2012	29,46667	25,83333	73	1008,2	1,866667	7,4	0
26/10/2012	28,13333	24,93333	75,66667	1008,933	1,633333	2,9	0
27/10/2012	28,3	25,33333	78	1010,767	1,366667	6	0
28/10/2012	27,8	25,33333	82	1010,167	0,833333	7,1	4,7
29/10/2012	28,93333	25,96667	78	1010,933	1,966667	9,3	3,7
30/10/2012	29,6	25,56667	71	1009,533	2	10,5	0,3
31/10/2012	29,33333	25,56667	72	1008,033	1,8	10	0
1/11/2012	29,96667	25,26667	67,66667	1008,267	1,9	9,6	0
2/11/2012	29,46667	26	75	1009,867	1,1	7,3	0
3/11/2012	29,3	25,4	71,66667	1011	2,066667	9,3	1,1
4/11/2012	29,8	25,83333	72,33333	1009,133	2,033333	10,5	0
5/11/2012	28,96667	26,3	80,66667	1008,1	1,566667	8,6	0
6/11/2012	29,06667	26,4	79,66667	1008,567	1,266667	7	12,8
7/11/2012	28,13333	25,66667	82,66667	1008,7	1,5	7,4	12,1
8/11/2012	29,3	27,85	75,5	1007,75	1,5	7,3	24,9
9/11/2012	29,06667	25,23333	72,33333	1009,1	1,9	9,1	0,4
10/11/2012	29,26667	25,66667	72	1007,2	2,166667	9,1	0
11/11/2012	29,73333	25,8	74,66667	1008,167	2,333333	7,1	0
12/11/2012	29,6	25,93333	74	1008,467	2,066667	9,2	0
13/11/2012	28,8	25,36667	75	1008,933	2,233333	0,9	0
14/11/2012	27,7	25,6	84	1008,833	1,7	7,6	0
15/11/2012	28,7	25,16667	75,33333	1008,1	1,833333	9,4	19,4

16/11/2012	29,23333	25,43333	72,33333	1008,633	2,466667	9,5	0
17/11/2012	29,23333	25,2	70,66667	1009,533	2,366667	8	0
18/11/2012	29,56667	25,4	70,33333	1009,933	2,233333	9,7	0
19/11/2012	29,33333	25,7	74,33333	1009,6	1,866667	5,8	0
20/11/2012	29,06667	26,26667	77,66667	1009,067	2	6,3	0
21/11/2012	29,26667	25,73333	74	1008,467	1,1	7,2	1,1
22/11/2012	29,7	25,83333	73,33333	1008,467	2,133333	9,3	0
23/11/2012	29,86667	25,96667	71,66667	1008,033	1,6	9,6	0
24/11/2012	28,83333	26,1	83	1009,233	1,133333	1,7	0
25/11/2012	26,86667	25,8	89,33333	1009,433	1,2	4	0,8
26/11/2012	29,2	25,93333	73,66667	1008,967	1,766667	9,2	27,8
27/11/2012	30,26667	26,1	69,33333	1008,367	1,7	9,9	0
28/11/2012	29,7	26,06667	75	1007,067	1,366667	8,1	0
29/11/2012	28,4	25,66667	83,33333	1007,967	1,166667	1,7	0
30/11/2012	26,83333	24,93333	84	1008,4	0,7	1,4	2,6
1/12/2012	28,36667	25,4	77,33333	1008,133	1,466667	6,1	0
2/12/2012	29,1	25,43333	76	1007,967	1,5	8,3	2,4
3/12/2012	29	25,8	79,66667	1009,367	1,266667	7,6	3,1
4/12/2012	28,36667	25,6	73,33333	1009,9	0,933333	7,8	3,6
5/12/2012	29,2	25,6	74	1009,167	1,233333	9	0
6/12/2012	29,36667	25,8	78,33333	1007,3	1,566667	5,5	0
7/12/2012	27,76667	26	82,66667	1007,767	0,666667	3,9	5,8
8/12/2012	29,56667	26,16667	73,66667	1009,8	1,866667	9,8	7,5
9/12/2012	30,03333	26,23333	77	1009,667	1,5	8,8	0
10/12/2012	28,86667	25,7	77,33333	1008,4	1,766667	8,7	2,1
11/12/2012	29,26667	26,16667	75,33333	1007,4	1,833333	7	6,9
12/12/2012	29,66667	26,06667	74	1007,533	1,966667	9,5	4,4
13/12/2012	29,86667	26,06667	73	1006,833	1,966667	8,9	0
14/12/2012	29,86667	26,26667	76,66667	1006,967	1,7	7,3	0
15/12/2012	28,96667	26,2	80,33333	1007,1	1,066667	6	0,7
16/12/2012	27,1	25,66667	90	1008,633	0,766667	0	4,5
17/12/2012	27,16667	25,3	84	1008,767	1,566667	4,1	9,8
18/12/2012	28,9	26,13333	79,33333	1008,2	0,933333	6,8	1,6
19/12/2012	29,16667	25,86667	79	1006,867	1,6	8,3	7,2
20/12/2012	26,7	25,36667	87,66667	1006,667	0,966667	4,6	41,9
21/12/2012	28,66667	26,26667	82,33333	1007,667	1,266667	7,4	2,8

yes

I want morebooks!

Buy your books fast and straightforward online - at one of world's fastest growing online book stores! Environmentally sound due to Print-on-Demand technologies.

Buy your books online at

www.morebooks.shop

Kaufen Sie Ihre Bücher schnell und unkompliziert online – auf einer der am schnellsten wachsenden Buchhandelsplattformen weltweit! Dank Print-On-Demand umwelt- und ressourcenschonend produzi ert.

Bücher schneller online kaufen

www.morebooks.shop

info@omniscriptum.com
www.omniscriptum.com

Printed by Books on Demand GmbH, Norderstedt / Germany